The METAMORPHOSIS OF PLANTS

Essays by

Jochen Bockemühl

and

Andreas Suchantke

Translated from the German and Introduced by Norman Skillen

NOVALIS PRESS

CAPE TOWN

NOVALIS EDUCATION SERIES

Typeset in 9.5pt Times Roman by Prototype Graphics & Documents, Cape Town.

Published by Novalis Press, P.O. Box 53090, Kenilworth 7745, Cape Town, South Africa.

Printed and bound in the Republic of South Africa by Mills Litho (Pty) Ltd, Cape Town.

Cover —
colour background: Sonja Shepherd.
sketch *Convolvulus althaeoides:* Andreas Suchantke.

"The Leaf: 'The True Proteus'" by Andreas Suchantke published as *Das Blatt – "der wahre Proteus"* in *die Drei*, Zeitschrift für Wissenschaft, Kunst und soziales Leben 6, June 1983. © 1983 Verlag Freies Geistesleben GmbH, Stuttgart.

"Morphic Movements in the Vegetative Leaves of Higher Plants" by Jochen Bockemühl published as *Bildbewegungen im Laubblattbereich höherer Pflanzen* in Goetheanistische Naturwissenschaft 2: Botanik, pp 17-35. © 1982 Verlag Freies Geistesleben GmbH, Stuttgart.

"The Morphic Movements of Plants as Expressions of the Temporal Body" by Jochen Bockemühl published as *Äußerungen des Zeitleibes in den Bildebewegungen der Pflanzen* in Goetheanistische Naturwissenschaft 2: Botanik, pp 36-43. © 1982 Verlag Freies Geistesleben GmbH, Stuttgart.

"The Metamorphosis of Plants as an Expression of Juvenilisation in the Process of Evolution" by Andreas Suchantke published as *Die Metamorphose der Pflanzen. Ausdruck von Verjüngenstendenzen in der Evolution* in *die Drei*, Zeitschrift für Anthroposophie 7/8, Jul/Aug 1990. ©1990 Verlag Freies Geistesleben GmbH, Stuttgart.

ISBN 0–9583885–2–0

CONTENTS

Acknowledgements —

The Novalis Institute would like to thank the Embassy of the Japanese Government in South Africa for their assistance in meeting the costs of publishing the Novalis Education Series.

Their generous contribution towards this project is an expression of the support of the Japanese Government for teacher education in South Africa.

The translation of these essays from the original German was kindly supported by the Section for Pedagogical Research at the Association of Waldorf Schools in Stuttgart, Germany.

Introduction

RATIONAL ORGANICISM

Norman Skillen

Some time in the late 1920s, during the heyday of theoretical physics, the great Cambridge researcher Rutherford, in a characteristic remark, said that "science is either physics or stamp-collecting". The attitude of mind behind such cocksureness goes back to the 18th century, for it was then that western ideas of nature were being forged. It can be said indeed, and this has been shown very clearly by Basil Willey[1], that Nature was the leading thought of the 18th century, writer after writer addressing the question of what constitutes natural behaviour, or natural religion, or natural knowledge.

In the area of knowledge, by the end of the century, a certain confused one-sidedness had emerged, and it was out of his impatience with this confusion that S.T. Coleridge, whom we must count as England's first major organicist, revived an old distinction in an attempt to clear it up. He remarked, in *The Statesman's Manual* [2]: "we have not yet attained to a science of nature", this being because the distinction between *natura naturata* and *natura naturans* was not properly understood, or, which was worse, not even recognised. Coleridge devoted considerable effort, indeed much of his life, to elucidating what this distinction means. The problem as he saw it was that the science of his day only recognised *natura naturata*, i.e. the world of sense objects, "nature in the passive sense", and so it expected phenomena to be derived from other phenomena. It had thus embarked upon a search for ultimate objects, whose properties, generalised into laws, would constitute knowledge. *Natura naturans*, "nature in the active sense", in process of becoming, or quite simply "life", which, Coleridge saw, is not phenomenal, cannot be grasped in the same way as *naturata*. *Natura naturans* requires penetration to the "inside" of nature, to the activity of which the phenomena are representations.

The great contribution he made in clarifying this situation, although you have to dig to find it (this has fortunately been done for us by Owen Barfield in his *What Coleridge Thought*), was to recognise that the apprehension of these two different "natures" corresponds to the action of two equally different mental operations. Both are aspects of reason, which for Coleridge was the "productive unity" out of which they arise. The mental operation corresponding to *natura naturate* he calls *understanding*, that corresponding to *natura naturans* he calls *imagination*.

Imagination holds a central place in Coleridge's philosophy, with regard to both poetry and science. On the one hand it is the basis of our being able to perceive anything at all (primary imagination), while on the other, to use Coleridge's own words "it dissolves, diffuses, dissipates, in order to recreate; or where this process is impossible, yet still, at all events, it struggles to idealise and to unify. It is essentially vital ..." (secondary imagination)[3]. By contrast, understanding represents the mind's separating, distinguishing tendency, it is "the faculty of judging according to sense", "the faculty of suiting measures to circumstances", "the aggregative power". It will be apparent that as a gesture towards the world, understanding will lead to abstraction, imagination to integration, participative identification. It is further maintained by Owen Barfield[4], that the basic act of imagination is the apprehension of polarity. Polarity is another key element in Coleridgean epistemology, and it is clear that to perceive two opposing tendencies as polarities, i.e. mutually *productive* rather than exclusive opposites, requires the operation of a mental power that "struggles to idealise and to unify".

Coleridge often promised that he would present a definitive work on his Dynamic Philosophy, but it never materialised, except as a vast array of fragments. Thus his high hopes for transforming English pragmatism into something like German "Naturphilosophie", which he had learned from reading, among others, Schelling and Goethe, came to nothing. Science in 19th century England stuck steadfastly to the path of understanding, looking always for objects of sense behind the objects of sense. Although J.S. Mill spoke favourably of the "Germano-Coleridgean doctrine" having turned 18th century thinking on its head, it had little influence. What a different theory of evolution we might have had from Darwin if Coleridge, and not Mill or Bentham, had been the leading philosopher of the 19th century. But the path of understanding was too obviously bearing technological fruit for anyone to worry about its ultimate philosophical validity, and this pattern continued relentlessly.

Right into the 20th century it was still possible for physicists to make remarks like the one I began with. Since Rutherford's time, however, and particularly since the 1960's, doubts, anomalies, alternatives, disaffections, not to speak of the environmental and military consequences of an "understanding-based" science, have been building up so that such cocksure arrogance is no longer possible.

This has come about, I believe, because understanding, pursued as the sole test of thought, eventually empties the world of meaning; and it seems that neither human beings, nor indeed the world itself, can long endure such meaninglessness. In our century its most extreme symptom has been the philosophy known as logical positivism. It sought to attack the claims of imagination by denying any meaning to the words in which they were made. To quote from an essay by Owen Barfield[5]: "The ground is cut away from beneath the feet of any idealist interpretation of the universe by a new dogma, not that such an interpretation is untrue but that it cannot even be advanced". That is, words (or descriptive models) with no directly perceptible or quantifiable physical referents are meaningless. That this includes words like "logical" and "positivism" does not seem to have bothered the proponents of this view, but the impossibility of its demands has been increasingly felt. Anomalies building up in evolutionary theory and research, as catalogued by Norman Macbeth[6], have only been outmatched by analogous developments in the realm of physics itself. And the burgeoning environmental awareness of recent years is based on a widely shared feeling for the intrinsic value and meaning of the natural world. The current plight of the natural environment is, as it were, its answer to meaninglessness, while scientific "heresy" seems to be the human world's answer, and together this represents a tremendous backlash, which seems to imply and anticipate a great change of world view. On the one hand environmental awareness is calling into question much that has long been taken for granted, while on the scientific side certain theories are now being taken seriously, which even 15 years ago would have been unthinkable.

Among these I would number: Gregory Bateson's theory of immanent mind, James Lovelock's "Gaia" theory of the Earth as a living organism, Rupert Sheldrake's theory of formative causation by morphic fields, and David Bohm's theory of implicate order[7]. Although each of these theories deals with a different area there has been considerable cross-fertilisation among them, and they all share a method which tries to proceed from the whole to the parts, or to develop a way of understanding how the whole can be contained or observed in the part. This, as we have already seen, requires applying the power of imagination directly to phenomena, and is precisely what was called for by Coleridge, and practised by Goethe two hundred years ago.

The previously mentioned essay by Barfield finishes with the injunction: "What is needed now is for someone to try the experiment of taking off his positivist spectacles and examining Goethe with the naked eye". This appeal was made in 1961. Fortunately the experiment had begun before then, and has gathered pace since, especially in Germany, and it is the purpose of this publication to place before an English-speaking public some of the fruits of this activity. Of course, the authors represented in this little collection have not just been "examining Goethe", although they have obviously done their homework in this regard, but have been assiduously applying his method. That certain developments in science, which are almost daily becoming less unorthodox, begin to converge with this approach is very timely and encouraging, but it should not belie the great gap which still exists between Goethean and "normal" science. I am inclined, although it is a considerable over-simplification, to paraphrase this as "between imagination and understanding"; for Goethean science is firmly based on applied imagination.

My own experience as a translator might serve to illustrate what this means. I had long realised that there was something in this idea of metamorphosis, but could never quite grasp it, in spite of attending courses and reading. The mistake I made, of course, was in thinking that metamorphosis is an idea *about* plants, whereas it is not an idea in this sense at all. It was only in translating these articles and being led, of necessity, inside the process of metamorphosis that my eyes were opened. Goethean thinking is all-too-easily misunderstood, as it is not a way of thinking about the world (though it also gives rise to such thoughts), but an active thinking *of* the world by means of applied imagination. This is what Goethe called *anschauende Urteilsskraft*, which has been translated, in the article by Suchantke where the phrase occurs, as "perceptual judgement". For the understanding this is a contradiction in terms, but not for imagination. "Perceptual judgement" quite literally expresses the polarity which is itself the span of imagination. On the one hand, imagination goes deeply into the activity of perception, while on the other it is linked to what Coleridge calls "the mind's self-experience in the act of thinking". These two poles are united in the act of applying imagination to the phenomena, so that it becomes the power by which we have the possibility, in our own thinking process, of perceiving the "thoughts" of nature. Gregory Bateson says we must learn to "think as nature thinks", and if we would know how to do this, I would say, we must look to Goethe and Coleridge, or to Bockemühl and Suchantke. In the following articles Bockemühl emphasises several times that this method requires much practice, and out of practising it himself he had developed, among other things, a language for describing transformations of leaf-shape. In the effort of translating this language I have been forced to apply it myself, with the result that I finally began to "understand", or arrive at a "perceptual judgement" of metamorphosis.

A chief effect of this has been to make supposedly ordinary things extraordinary — the archetype is absolutely "democratic" and extends its formative influence even to the dustiest plant by the side of road. Conveying the full import of the wonder than can be aroused by such apparently lowly phenomena is one of Goetheanism's chief difficulties. What I mean here is illustrated rather well by a passage in a story by A.A. Milne. Pooh and Piglet have just been to visit Rabbit, and have been ticked off for wasting his time, and as they walk away we get the following conversation:

"Rabbit's clever," said Pooh thoughtfully.

"Yes," said Piglet, "Rabbit's clever."

"And he has Brain."

"Yes," said Piglet, "Rabbit has Brain".

There was a long silence.

"I suppose," said Pooh, "that's why he never understands anything"

Rabbit's problem is, or course, that he "understands" too much. He is too much on the look-out for the significant detail to be able to appreciate the awesome mundanity of being wished a "happy Thursday", which was why Pooh and Piglet had visited him in the first place. Finding a balance between imaginative involvement and articulate analysis is not very easy when one is dealing with "open secrets" (like Thursday), which was Goethe's term for such phenomena as the metamorphosis of plants. On this basis we can appreciate what Bockemühl and Suchantke have achieved in getting their "imagined judgements" into words at all. We must also appreciate that the method they have used requires an epistemology which follows the criteria of such principles as analogy and correspondence, as well as homology and heredity. The articles themselves are a practical demonstration of the workings of such principles.

To finish, I would like to point out one further dimension of "Goethean thinking". As well as being scientific, it is also poetic. How could it be otherwise for a discipline based on applied imagination? But far from detracting from its scientific status, the far-reaching significance of this is that it opens up the possibility of extending, or restoring, knowledge-value to poetry (and by implication to art). Once again Coleridge has already covered this ground. In the *Statesman's Manual* [8] he gives an electrifying definition of the symbol, which finishes with the words: "A symbol always partakes of the reality which it renders intelligible; and while it enunciates the whole, abides itself as a living part in that Unity, of which it is the representative". Coleridge is speaking here, of course, of the poetic symbol, but it is by a similar path of reasoning, starting from the scientific direction, that Andreas Suchantke, in the final article in this collection, comes to speak of the Rose and the Lotus Flower as "natural symbols". It would seem that between archetypal ideas and poetic symbols, understood in Coleridge's sense, there is no essential difference.

Nevertheless, Goethean thinking remains scientific. Quite apart from its other credentials in this regard, it displays a chief feature of scientific theories in that it has predictive power. In the articles that follow it can be seen very clearly that predictions made by Bockemühl and taken up by Suchantke, yield astounding results. For this reason it is advisable to read the articles in the order in which they stand. In this way the process of discovery can be followed, the groundwork having been laid by Suchantke in the first article. These discoveries, in turn, have very wide-ranging implications, which have either been dealt with elsewhere, as the references show, or await further investigation. Let it finally be emphasised once more, that such investigations, and the evaluation of their results, require more than the test of understanding alone (though they certainly require that as well). In other words, reading and thinking about this book is not enough. Its true significance will only be realised by a continually renewed act of imagination, which is a direct relation to the real world, and the only sound basis upon which to construct a science of *natura naturans*, a truly rational organicism.

NOTES :

1 Basil Willey: *The 18th Century Background*, Ark, London 1986.
2 Quoted in Owen Barfield: *What Coleridge Thought*, p.24.
3 S.T. Coleridge: *Biographia Literaria*, p.167, Dent, London 1980.
4 Owen Barfield: *What Coleridge Thought*, p.36, Wesleyan, Conn. 1978.
5 Owen Barfield: *The Rediscovery of Meaning* in book of same name, p.13, Wesleyan, Conn. 1977.
6 Norman Macbeth: *Darwin Retried.*
7 Gregory Bateson: *Mind and Nature*, Wildwood House, London 1979.
James Lovelock: *The Ages of Gaia*, OUP, 1989.
Rupert Sheldrake: *The Presence of the Past*, Fontana, London 1989.
David Bohm: *Wholeness and the Implicate Order*, RKP, 1981.
8 Kathleeen Coburn (ed.): *Inquiring Spirit*, p.103, Univ. of Toronto, 1979.

1

THE LEAF: "THE TRUE PROTEUS"

Andreas Suchantke

The two hundred years which have elapsed since Goethe first experimentally investigated and described the "metamorphosis of plants" have done nothing to make his whole approach to science more accessible. Over the same period science has followed a path of development so different from Goethe's methods that bridging the gap thus opened up is now well-nigh impossible. These methods remain shrouded in general incomprehension, and their instigator is dismissed as a dilettante, who lacked the basic knowledge of his subject, the essential details of which scientific investigation has since filled in. True, it is accepted that he provided science with some useful leads, but that same science has proved him wrong in his most important conclusions (colour theory, vertebral theory of the skull). In what follows, an attempt has been made to show why the strongly one-sided modern critical attitude cannot but regard Goethe as irrelevant, and how his form of thinking only reveals its true potential if we are prepared to overcome certain fixed ideas. These we will have absorbed through the all-pervasive influence of the scientific attitude in education. Once we have freed ourselves of this built-in bias however, Goethean thinking is capable of providing extraordinarily fruitful insights, leading to a deeper understanding of the plant world.

On his way back from Sicily, filled with impressions of a plant world rich in variety of form, Goethe writes to Herder on 17th May 1787 from Naples: "I must further confide in you that I am close to the secret of plant origin and organisation, and that it is the simplest thing you can imagine. Under heaven there are the most beautiful things waiting to be observed. The archetypal plant is turning out to be the most extraordinary thing in all creation for which nature herself might well envy me. With this model, and the key to it, there is simply no end to the plants you can invent, and they cannot but be consistent; that is to say, even if they do not exist, they could exist, and not like the shadowy fancies of painter or poet, but with an inherent truth and necessity. The same law will be applicable to all living things. But let me briefly state the point which will render everything more comprehensible to you. I had come to the realisation that in that organ we are wont to call leaf *lies the true Proteus, which can conceal and reveal itself in all plant structures. From top to bottom the plant is all leaf ..."* [1]

In the *Metamorphosis of Plants* Goethe then sets out in exemplary fashion how the organs of the blossom, the parts of the fruiting body, the husks and shells of seeds are all modified leaf structures. These he offers as convincing evidence of the "ultimate identity of all the parts of the plant". Really all? Stem and roots as well? For Goethe in the initial euphoria of his discovery there was no doubt. "Everything is leaf! And this very simplicity makes possible the greatest diversity. A leaf which merely sucks up moisture in the soil is called a root. A leaf swollen by moisture — a bulb. A leaf that expands lengthways — a stem or stalk." [2]

A grandiose generalisation, but one which surely casts its net much too widely! The central axis of a plant, the stem, is far more than a leaf. If anything, then it must be the true Proteus. All other organs grow from it, not least among them the various forms of leaves, and even roots can arise from the same source. Compared with this the leaf is a specialised structure which, once successfully formed, is not capable of producing anything further (this is the general rule in the plant kingdom, but where embryonic tissue remains around the edge of the leaf as in Bryophyllum, new plants can emerge vegetatively from it). Goethe came close to realising the central importance of the stem, or more exactly of the node from which the leaves, branches, and in some cases adventitious roots arise. "From node to node the plant essentially comes full circle; just as at the seed-stage it only requires a root-tip, a root node or a cotyledon node, so it only takes a sequence of nodes to give rise to another complete, viable plant." [3] The true Proteus, it would appear, is the node.

And what of the root? It is no modified leaf, but a structure *sui generis* which is subject to completely different physiological and morphological requirements. It contrasts strongly with the leaf in the position of its growing point. Whereas in the leaf this is at the base, in the root, corresponding to the meristem above ground, it is at the tip. The root has no relation to light unlike leaf, shoot and blossom; the main root even grows away from the light (negative geotropism). Shoot and root are mutually antagonistic, a polarity which appears even in the embryo. In his *Vorarbeiten zu einer Morphologie der Pflanzen* Goethe briefly outlines this polarity. Following a draft chapter on "Organic Unity" comes one entitled "Organic Duality" *. Unfortunately we cannot quote the entire text here — the reader may

* Translator's Footnote: Duality – the word used by Goethe is *Entzweigung*, which very concretely captures the process of splitting or branching or dividing which is meant. The German thus demands of English something like "dualisation", and it was thought better to point this out

consult it himself, and it is well worth doing so[4]. With consummate clarity Goethe unfolds in imagination the development from the seed of the whole plant organism:

> "Ideal unity: if these various parts are thought of as having arisen from an ideal *prima materia* and then as having passed through a series of formative stages, this ideal *prima materia*, conceived at its very simplest must be thought of as intrinsically dual in nature; for without envisaging such prior duality it is impossible to imagine a third entity arising from it ... wellspring of both root and leaf. They are originally united; indeed the one is unthinkable without the other. They are also originally opposed to each other. We answer the question of why the root tip grows downwards and the dicotyledons upwards by saying that they are a specific instance of the general polarised dualism of nature. With this insight we can go on to say something of the particular conditions for plant life. A plant, like any natural entity, cannot be thought of in isolation from the conditions of its surroundings. It requires a substrate providing support and basic nutrients adequate to its size. It requires the more rarefied nutrient source of air and light to promote diversity of structure. We find that the root requires damp and gloom, the leaves light and dryness in order to develop. Thus from start to finish these requirements are polarised ..."

Comparing this draft with the final version of the *Metamorphosis of Plants* reveals a radical change in the picture presented — instead of the whole body, a torso, as it were. For Goethe confines himself to consideration of only one of the two poles, the part above ground; and of this he selects for attention mainly the changes of shape its leaves undergo. It has become, in effect, a "Metamorphosis of the Leaf".

The charge that he neglected the roots was of course often levelled at Goethe. In 1824 he offered an indignant counterblast to this "Unfair Claim" as his short reply was entitled [5]. By means of an image he tried to get things back into proportion. "I have as much respect for the root as for the foundations of the cathedrals of Strasbourg and Cologne, and their construction is not wholly unfamiliar to me ... but our actual contemplation of the building begins with the surface exterior." This not entirely happy comparison speaks for itself and reveals the significance which the root has in Goethe's eyes. He finally comes out with it himself as clearly as could be desired in the last paragraph.

> "The root — it really was none of my business; for what had I to do with a structure which limits itself to filaments, strings, balls and knots, and that in no edifying order; which in spite of endless twists and turns shows no sign of the process of intensification necessary for the occur-

---

rather than use the word. "Duality" must therefore serve, even though it refers more to a state than a process. This note should be borne in mind at the other places where "duality" occurs in the text.

rence of transformation*. It is the pursuit of this process alone which has dictated and kept me to my chosen path. Let everyone follow his own line at his own pace, and in forty years look back upon what he has achieved which, according to a certain man of genius, is all we can hope for."

They were quite simply offensive to this eye-centred aesthete, these tangles of roots. He could get nothing from their formlessness. But these utterances should not just be glossed over, for Goethe indeed revealed in these few lines much of what preoccupied him most deeply in his research. He was convinced that in metamorphosis he had found a universal principle which was in no way restricted to plants. Occasional remarks in conversation indicate this. "It is always just this same metamorphosis or the ability in nature to effect transformation which produces a flower — a rose — from a leaf, a caterpillar from an egg, and a butterfly from a caterpillar" he remarked to Falk in 1813, and two years later Suplice Boisseree records Goethe's opinion that "Everything in life is metamorphosis, among plants and among animals, even up to mankind as well." And then, in Boisseree's always somewhat slap-dash style "Everything is so simple, and always the same; there is truly no art in being the Lord God, once creation is there *it contains only one thought ...*" [6]

There is absolutely nothing of this to be found in the root which presents itself in "endless twists and turns" in "no edifying order", with "no sign of the process of intensification. It is merely an aggregate continually added to with no structural diversity." Unlike the leaf which "shows the highest degree of diversity in its development and by stages approaches perfection." ("Organic Duality".)

The root is formless, or at any rate the most morphologically deficient part of the plant. Goethe, who once testified in a letter to Alexander von Humboldt that he "always proceeded from the form as a whole" ** — in contrast to von Humboldt, who always took the part, the single element as his starting point — found that the root offered nothing to the student of coherent form. A plant's changing sequence of leaves, their modifications and transformations, in short, their "metamorphosis" is for Goethe emblematic of the working of an underlying formative power emanating from an ideal realm. In the sense-perceptible composite form of a plant the ideal becomes actual, for the visible body of the plant results from the activity of the archetypal plant. In contrast, the root is not only lacking in form, but also removed from sight. It is literally unobservable. Dig it up and it falls apart, the ends all torn off. Cultivate it in a narrow, vertical glass beaker and it will be rendered visible, but only in two dimensions. All attempts

* Translator's Footnote: This long phrase: "the process of intensification necessary for the occurence of transformation" is a rendering of the German word *Steigerung*. By this Goethe meant the increase of a morphological tendency, e.g. contraction, to a point out of a morphological tendency, e.g. contraction, to a point out of which the next transformatory stage then emerges.

** Translator's Footnote: "the form as a whole" : the German word here is *Gestalt*.

to make the root observable involve interference of some sort so that the pure, unadulterated phenomenon remains elusive. *The root can be reconstructed mentally, but is not accessible to primary sense experience.* Only as a specialised structure, the sort of thing found chiefly in the tropics, can the root be directly observed, e.g. serial roots. But then they often take on the character of normally visible parts of the plant like the trunk, as in the Strangler Fig, and in the holy Banyan Tree; or else they flatten out, turn green, and take over the assimilatory function of leaves, as in certain orchids.

And yet — if Goethe's approach, his method of research did not permit access to the root, then it would indeed be of extremely limited significance; for what use is the concept of the archetypal plant if its perceptible counterpart is only a "torso"?

That this is not the case is shown by the previously quoted "Draft", which takes account of the *whole* plant. It is the further elaboration of this which ends up as a "torso" (which in no way detracts from the total achievement). The reason for this lies not in the method but as has been shown, in Goethe's turn of mind, and in that which for him took priority and interested him most profoundly. Rudolf Steiner described the situation in the following terms: "Goethe's work does not constitute a thoroughgoing account of nature in its entirety, but only fragments of such an account. He who wishes to gain a complete picture of the world of Goethe's ideas on this subject will have to fill in the gaps for himself." [7] Let us see therefore — not presumptuously, but taking Goethe's lead, i.e. with the help of his own method — *whether via the leaf, this "true Proteus", we can indeed approach an understanding of the root.* Before proceeding however, a certain point needs clearing up, and this requires a short digression.

In the middle of the last century, the English anatomist Owen established a conceptual framework with far-reaching consequences. It was intended to reduce morphology to the status of an auxiliary science in the service of botanical and zoological taxonomy and above all, of phylogenetic classification[8]. Faced with organs displaying structural similarities, he distinguished between those whose similarity could be traced back to *the same physical antecedents,* and those which, although strikingly similar in form, were constructed from *divergent material antecedents.* Ever since Owen, the former have been called *homologies*, the latter *analogies.* Homologous organs need not be of similar appearance — quite the contrary. Only after thorough investigation of the skeleton, embryonic development etc, can it be demonstrated that such different structures as birds' wings, whales' fins, indeed all vertebrate forelimbs, are homologous organs. The same goes for structures so different in function as the hammer and anvil of the human middle ear and the (so-called) primary jaw-joint of amphibians and reptiles. In contrast, bird and insect wings are analogous structures — the wings of butterflies are not modified forelimbs, but outgrowths of the dorsal skin. Equally analogous are the widespread "cactoid" growth-forms occurring in species of Spurge, Carrion

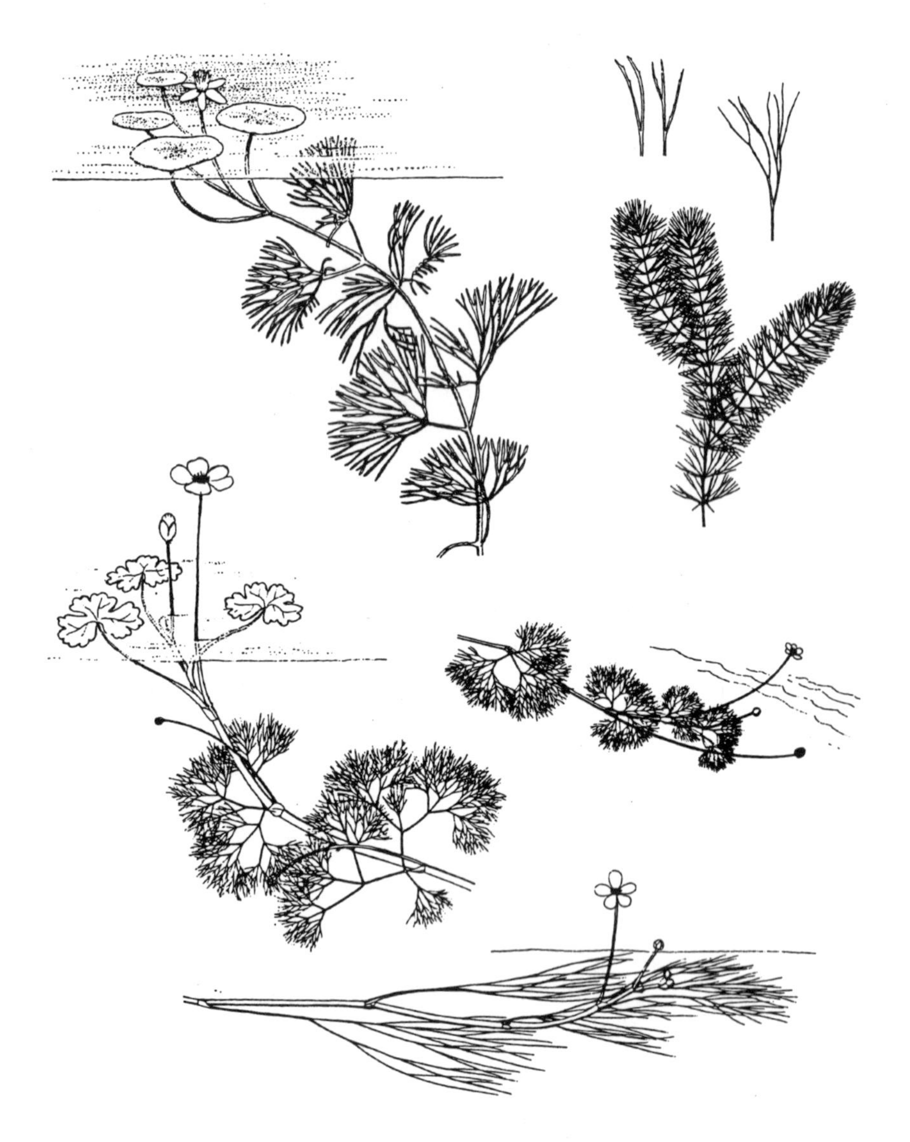

Above — two relatives of the Water-Lily: left, *Cabomba* with leaves plate-like on the surface and dissected under water. Right, the Hornwort *Ceratophyllum demersum* which only has finely dissected underwater leaves; above it enlarged leaf-segments of the two European species *Ceratophyllum demersum* (left) and *submersum* (right). *Below* — similar leaf structures in spp. of Water Crowfoot: left, *Ranunculus peltatus* resembling *Cabomba* with its two leaf forms, right, *Ranunculus baudotii*, and at the very bottom, *Ranunculus fluitans*: like *Ceratophyllum* both have only hair-like, dissected, underwater leaves. (Partly after Hess, Landolt, Hirzel.)

Flower, Vine (*Cissus*) and many more besides, but chiefly in the cacti themselves. Again and again the same external pattern appears in plants which, although not otherwise related, all occur in arid regions.

In research almost exclusive priority has been given to homology. Attention has centred upon tracing diverse phenomena (bird's wing, horse's hoof, whale's fin) to common hereditary aspects, thus enabling reconstruction of a corporeal succession, the "family tree" so to speak. Analogies were of no interest and were dismissed as "adaptive similarities" or regarded as "vexations to truth" [9] disguising the real connections under a "veil of deception". [10]

Having thus evaluated the two principles and opted for homology in preference to analogy, science was set clearly upon an irreversible course towards a materialistic biology. This led consistently away from Goethe's approach. Thus his "vertebral theory of the skull" mentioned earlier has long been regarded as totally refuted. The truth of the matter is that Goethe has not been properly understood. He was not trying to prove that the bones of the skull are homologous with the vertebrae, in other words, that they originate from the same embryonic tissue. He saw in them in modified, metamorphosed form the workings of the same formative principles as in the vertebrae.

Goethe was not interested in the physical origin of this or that organ, or this or that plant. He was looking for the concrete ideas, which, regardless of its hereditary source, imbue organic material with form and thus come to expression. We have every right to assume that he would have been a keen upholder of Owen's conceptual framework, and, while fully recognising the criteria of homology would also have drawn attention to the significance of analogies. They reveal the activity of some formative principle which is not derivable from the material nature of the organism[11].

This Goethean position does not preclude the distinction between homology and analogy, but rather *transcends* it. Let us then assume this stance and try to discern what the language of plant form, regardless of the organic or hereditary origin of the structures involved, actually has to tell us. Using this approach, it is forcefully brought home to us that the essential nature of the plant is in no other organ so clearly and comprehensively expressed as in the *leaf*. It is mainly surface, and therein lies a basic feature of the plant (including the root) namely that it is all "outside". Its inside is not only appreciably simple in structure, even when compared to the inner organisation of relatively primitive and lower animals; it is also negligible in purely quantitative terms in relation to its vast surface area. This brings out the contrast between plant organisation and that of the animal, which has tended in the course of evolution towards increased interior complexity and differentiation, in parallel to an increasingly streamlined exterior.

This contrast, or rather polarity between plant and, above all, higher animal is also shown by the fact that the plant surface is not a barrier, closing the bodily interior off from the environment by means of scales, fur or feathers, but a transparent filter and permeable membrane in and on which substances interact in a variety of processes. These substances come partly from the surrounding atmos-

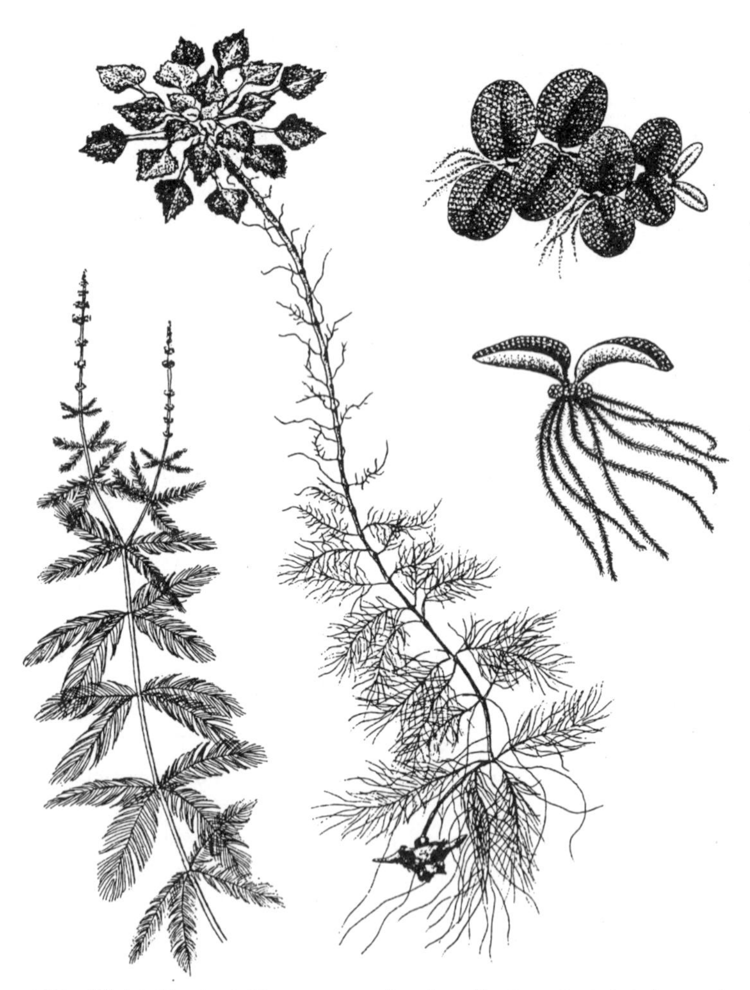

In the middle, Water Chestnut *(Trapa natans)*, a free-floating plant that does not root in the soil: the rosette at the top lies on the surface of the water, while at the other end the seed from which the plant germinated is still attached. As the stem grows and the leaves it first puts out wilt and fall off they are replaced at each node by finely divided structures which, although green, are genuine roots (after Chodat). *Left*, the Spiked Water Milfoil (*Myriophyllum specatum*) the uniform structures of which are all underwater leaves (only those near the blossom project out of the water). This plant belongs to the *Haloragaceae* Family, in which broad surface-leaves and finely dissected underwater leaves are never found in one and the same species, but always confined to different genera: the species of *Gunnera* that occur in the wet, misty forests of the Andean slopes, in the south of S. America, and are often cultivated in botanic gardens have gigantic leaves, reminiscent of Rhubarb, which can be as much as six metres wide. *Above right*, patrolling the water's surface, a Salvinia (*Salvinia* sp.), with five pairs of fronds, of which the last is still in process of forming. *Below right*, in side-view, a section of shoot, with two leaf-like fronds and a third in the form of strands covered in delicate hairs and hanging down in the water like roots. The structures in the middle are spore-capsules.

phere, the realm of air and light, and partly from the water and mineral spheres of the soil, both inorganic sources. At this metabolic interface they are then integrated into a living context, and lifted in the process up to the level of organic life. Processes of condensation or solid formation mingle with dissolving or fluid state processes. On the one hand glucose is produced from gaseous and liquid components and fixed in starch and cellulose, while on the other hand, dissolved mineral salts become incorporated into the streaming movements of the cytoplasm.

Condensation and dissolving, "contraction and expansion" to make use of the pair of concepts put forward by Goethe, and understood by him always in a flexible and dynamic sense: "Through the sequence of changes the parts of the plant undergo, a force is at work which I can only rather awkwardly describe as expansion and contraction. It would be better to use algebraic symbols like x or y, since the words expansion and contraction do not express the full range of effects produced by this activity. It contracts, expands, constructs, modifies, permeates, withdraws, and only by seeing these multifarious effects as a *unity* can we attain a clear perception of what this panoply of words was intended to illustrate and explain ..." [12]

Does not everything about the leaf mark it as the prime expression of these processes of interpenetrating and mutually productive polarity? Expansion appears in the broad and rounded leaf-blade which will often be sub-divided into points, leaflets and lobes, all peripherally extended. It is also evident in the branching and fanning out of the leaf-veins, while contraction features in their re-uniting to form the anastomoses, in their joining up to form the mid-rib, and in their bunching together in the stalk. To go from the edge to the centre of the leaf and from the tip towards stalk and stem is to follow the descending stream of sap and thereby participate in the gesture of contraction. The opposite course, following the ascending sap-stream with its cargo of dissolved mineral salts from the stalk to the leaf-blade with its spreading veins, is the process of expansion.

Thus the leaf is a perfectly balanced, composite expression of that which appears at the opposite poles of the plant in a thoroughly one-sided form. In the root is mainly found in the all-round absorption of water and salts, the contractive tendency encountered before in the veins and stalks of the leaves while the blossom, giving off its scent and scattering pollen, is all expansion. It must be admitted of course that in both cases the contrary tendencies are also present, but in form rather than in function. The roots grow centrifugally, spreading out on all sides, and the closer structures come to the blossom, and more particularly to the fruit, the more "contracted" they are. If we recall the passage by Goethe just quoted, it continues as if tailor-made for this state of affairs as indeed it is. "To gain knowledge, man ... must divide that which is not to be divided; and in this case there is no other option but to *re-unite* that which nature has presented to our cognition in divided form ..." Even if far more differentiated and structurally specialised, there is ultimately nothing at either end of the plant which is so fundamentally different in organisation or totally new, that was not already present in the leaf.

It is worth underlining here that the root is not in some way homologous with the leaf, and is not, according to homological criteria, a metamorphosed leaf. But the leaf does possess structural elements and physiological processes which, in intensified and specialised form, are also found in the root. The root also embellishes these with elements not found in the leaf — at least not in general. Nevertheless, there are cases in which the leaves adopt the function of the root, as in the rootless epiphytes and desert-dwelling Bromelias (Pineapple family), whose leaves have special scales for the uptake of wind-blown mineral substances. Thus the leaf can when necessary take the place of the root.

The leaf however is much more than this. It is really the "true Proteus" being not only the mid-point at which polarities interpenetrate, but also the source from which they originate. *The leaf goes back to the very origin of plant-life, and is the form in which the plant first appears.* This assertion, which to many might seem rather surprising — the planiform leaf is after all regarded as a late achievement in the evolution of land plants — can be easily justified by recalling the function of the leaf. It is a permeable boundary layer where the substances and energies of the twin realms of atmosphere and soil interact in living process and are thus integrated into a higher third entity, the plant as a composite whole. This can be observed going on today in the most primitive of all plants, the Algae. By this is not meant those species that grow in coastal areas where the presence of three-dimensional space in the form of cliffs and rocks jutting out into the sea facilitates growths that in their apparent differentiations are so reminiscent of land plants. No, we are referring here to the most simple of Algae, which *cover the open sea with a delicate green skin, a boundary layer over the surface of the water.* They are termed unicellular Algae, but this does not do justice to them as a living phenomenon. As unicells they are humanised artefacts of the microscope, abstracted from their natural context and enlarged in visual isolation. In undisturbed real life the single cell is meaningless, functionally as well as visually — *the totality, the layer as a whole is what counts.*

Plant life originates as a boundary layer where earth and cosmos impinge upon each other. This is the *archetypal leaf*, scarcely yet in a solid state, living with no trace of contraction in a condition of complete expansion. And this earliest form of plant existence is so prototypical — or Protean — that in the further development of the plant kingdom it is retained as the formative and functional motif to which all others are subordinated.

Significantly, this archetypal leaf-form later appears at its purest among those higher flowering plants which have once more become aquatic, and have leaves that spread out on the surface of the water. In these plants the leaf-form comes as close as is possible to the perfection and simplicity of the circle, thus contracted to a small-scale image of the earth-spanning expanse of the oceans. Many otherwise unrelated flowering plants conform to this pattern, the Water-Lilies leading the field. In addition to them there are the tiny duckweeds, the Frog Bit (*Hydrocharis morsus-ranae*, which in French, significantly enough, is called Petit Nenuphar, the small Water-Lily), many species of Pondweed (*Potamogeton*), the golden yellow-flowering Fringed Water-Lily (*Nymphoides peltata*) related to Bogbean and Gentian. Even ferns figure among this company; the Kariba Weed (*Salvinia*) with its oval-shaped fronds, or the Water Clover (*Marsilia*) with its remarkable four "leaflets", which reach up on their long, thin stalks and group themselves in a circle on the water's surface.

The clear perfection of outline displayed by these leaves is intensified by their inherent polarity, i.e., the way the round, spread-out leaf-blade is so sharply distinguished from the long, tight, slender stalk — circle and radius. The latter is always submerged. *Significantly, leaves which live submerged also take on this long, thin, filamentous form.* Practically nothing remains but the leaf-ribs, and the water is filled with wafting strands which resemble roots even if green. We do not have to look far to find, in a number of related groups, species with both sorts of leaves. The submerged sort dissected into filaments, and the round sort that spreads out on the surface. Many species of Water Crowfoot, e.g. *Ranunculus aquitalis* and *peltatus*, and Cabomba, a relative of the Water Lily, display such differently-formed floating and subaqueous leaves. The same is found in the already mentioned Kariba Weed (*Salvinia*). At every node on the shoot three fronds are formed. Of these, two are Water-Lily-like and cling to the surface, while the third is dispersed in a series of root-like strands which hang down in the water. No one seeing them would get the idea he was looking at leaf *homologues*, their *analogous* resemblance to roots being so great. In the aquatic crowfoot and water-lily groups there are also other species which only have submerged leaves, which are all dissected into filaments as would be expected. Among the water-lilies such are the Hornwort species (*Ceratophyllym*), while water crowfoot species with this feature are *Ranunculus fluitans, trichophyllus, baudotii* and others.

The picture so far built up becomes even more instructive if a further plant following the same pattern is added into it. The relatively broad floating leaves show the same tendency to roundness and arrange themselves in a rosette-like mosaic, while under water the same finely dissected forms are found, but in this case, although green, these are genuine roots. This is the Water Chestnut (*Trapa natans*). Where it has real roots, Water Crowfoot, *Cabomba* etc. have very similar underwater organs, which are leaves analogous to roots. The connection between these two genuine root and root-like leaf, is life in the mineral-rich hydrosphere. *In this region the leaf becomes root-like.* Its persistent greenness is accounted for, as is that of the Water Chestnut roots in the same environment, by light penetrating the upper reaches of the water.

The opposite course is taken by one of the Water-Lily's close relatives, the Lotus Flower (*Nelumbo nucifera*), which only spreads out some of its leaves on the water's surface and from time to time lifts single ones high up out of the water on long stalks, just as it does with its large, radiant blossoms. Unlike the flat, plate-like leaves lying on the water, those up above it, although just as round, are bowl or chalice-shaped by virtue of their billowing edges. The same appears in the highly raised receptacle, which in the Lotus Flower grows so far up around the fruits, that these look as if they have sunk deep into it. The fruits of the "root-leaved" Hornwort (*Ceratophyllum*), on the other hand, tower one by one high above it.

These are two very telling gestures. The almost chalice-shaped Lotus leaves and the dissected, underwater leaves of *Cabomba* and *Ceratophyllum* have diverged in opposing directions from the central position occupied by the Water-Lily. Both tendencies indicate the steps taken by the plant world as a whole on its evolutionary path from water to land, from archetypal leaf at the most primitive algal level to highly differentiated flowering plant — from "unity" to "organic duality".

This path meant breaking through to three-dimensional space from being pinioned in the plane where the water and mineral sphere meets that of light and air. On land these two bordering spheres are found to have changed in character, drawn further apart so to speak: the upper one dryer, the lower one more solid and, above all, devoid of light. How does the plant respond to this? If at all, then only in close proximity to the interface does it produce leaves with a large surface area, among which broad, rounded basal leaves and rosettes are clearly echoes of the archetypal leaf. The higher the plant grows the more restraint it shows, tending inexorably towards the state in which all its above-ground structures begin, that of the bud. The vegetative leaves still unfold their surfaces around the stem preserving distance from each other, while the petals are grouped more closely together and display a clear tendency towards a bowl or jug shape (which is also evident in the raised leaves of the Lotus Flower). This then culminates in the spherical structure of the fruit, the outer elements of which, the carpels or fruit-leaves, cleave more or less to the bud stage until, when ripe, they open. Thus, to the light streaming in on all sides the plant responds with encircling, ever more spherical structures.

In the cracks and fissures of the earth structures other than thin, string-like strands are unthinkable, at least as long as growth is still proceeding (dormant tubers and corms are something else); indeed, shoots heading upwards behave in a similar way. Once germinated, before they reach the light, they grow "like roots", in other words almost exclusively lengthways, with large gaps between nodes. As soon as they reach the light this form of growth slows down abruptly, and they begin the business of spreading their leaves. The point here is that the roots display to an intensified degree what is found also in the underwater leaves of Water Crowfoot and *Cabomba* etc., although not removed so far from the realm of air and light, filamentous and string-like structures. The root is in its radial structure the answer to the spherical form of the earth, into which, in obedience to gravity, every primary root grows.

Thus we can begin to understand the plant as the outcome of a dialogue between earth and cosmos. And as a three-fold organism, with two poles, one represented by the root, the other by the blossom and fruit, and in between the vegetative realm. This middle realm — the leaf — still has, as boundary layer and metabolic exchange surface, the character of the most primitive form of plant life.

The leaf, we must agree with Goethe, is the “true Proteus”. From top to bottom the plant is all leaf!

NOTES :

1 Italics by the author; quoted from the Insel Edition (Leipzig), vol. 4, p.399.

2 From *Goethes Sämtliche Werke*, vol. 16 (Naturwissensch. Schriften vol. 1), p.633, Insel Edition, Leipzig o.J. See also: *Goethes Morphologische Schriften*, selected and introduced by Wilhelm Troll, Jena 1926, p.52.

3 *Vorarbeiten zu einer Morphologie der Pflanzen*. Hand-written manuscript, 1788, 1789. Otherwise as in note 2.

4 This part is also included in the edition of *Goethes Naturwissenschaftlichen Schriften* dealt with by Rudolf Steiner (as an Annexure in Vol. 4 of *der Farbenlehre*).

5 Vol. 16 of the Insel edition of *Goethes Sämtliche Werke*, p.316.

6 Goethes Gespräche (minus the Gespräche mit Eckermann), ed. F. Frh. von Biedermann. Insel Edition 1957, p.281 & 332; italics by the author.

7 In *Goethes Naturwissenschaftliche Schriften* ed. Rudolf Steiner, Stuttgart/Berlin/Leipzig o.J., Vol. IV, 2. Footnote p.547.

8 Full account to be found in A. Remane: *Die Grundlagen des Natürlichen Systems, der Vergleichenden Anatomie und der Phylogenetik*. Leipzig 1952.

9 A. Remane, *op. cit.*

10 W. Troll: *Organisation und Gestalt im Bereich der Blüte*. Berlin 1928.

11 Nowadays, of course, this is explained as the interplay of natural selection and chance mutation, and traced back in the normal way to purely material processes. To go further into this, however important it might be, would lead too far away from the context of this essay. Nevertheless, the significance of deciding for homology while discarding all but physical causes becomes clear when we consider that morphology — understood, in such terms, as comparative anatomy — includes the human being in its investigations as a matter of course. Man thus becomes the object of a science based solely upon heredity, which has no way of dealing with the “other pole”, the spiritual dimension of man’s being which permeates the physical, because it simply cannot see it, having become blind to such a possibility.

12 *Vorarbeiten zu einer Morphologie der Pflanzen*. Hand-written manuscript 1788, 1789. Otherwise as in notes 2 & 3.

2

MORPHIC MOVEMENTS IN THE VEGETATIVE LEAVES OF HIGHER PLANTS

Jochen Bockemühl

For a study of the growth of plants two alternative lines of approach initially present themselves. One approach will discover in it a very intricate process, which, under increasingly precise investigation, breaks down into an ever-increasing number of active components. The other approach, through perceiving the process in its totality, will find itself predisposed to seek some higher unity behind it. In the effort to find this unity the former approach can then complement the latter in that the components, once singled out, can act as pointers. The more exactly they are studied, the clearer a picture can be formed of the unity: the archetype.

The present investigation has addressed itself to the task of tracing the workings of the archetype at the level of "movement". A few points of method required for this have been previously described (Bockemühl 1982). I will make direct use of these, and having considered a range of specific phenomena will put forward some ideas about the larger context to which I feel this work applies.

In trying to get closer to the archetype it is always important to select for attention a section of the organism which can be regarded as representative of the whole. The full series of mature vegetative leaves (leaf-sequence) of an annual plant, as was shown in the aforementioned paper, is identifiable as a single, self-contained movement. This in turn, as an integral part of the plant, reflects in a special way, the developmental movement as a whole.

Without detracting from this it must, however, be pointed out that the method

Figure 1

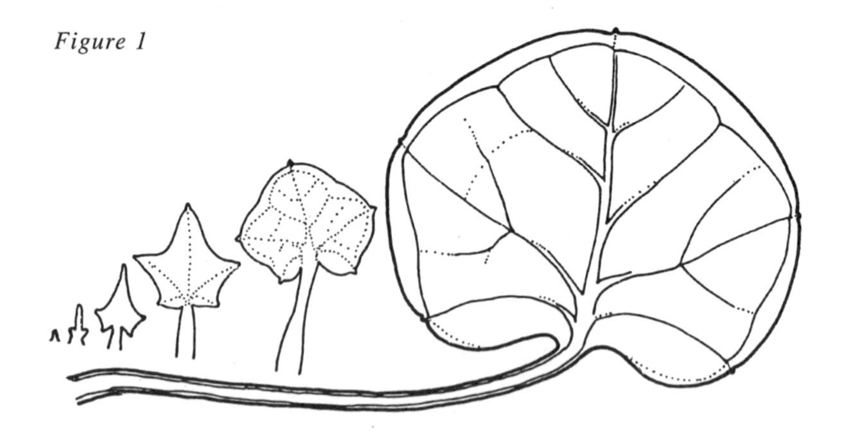

involved here, by which a mental picture is formed of one growth stage and then inwardly transformed into the next stage so that the transformation appears in imagination as a movement, can be just as fruitfully applied to the growth of individual structures as to that of whole organisms. I propose now to follow Splechtner (1930) in calling both *morphic movements* *.

1 *Morphic Movements in Individual Leaf Development*

We begin with the morphic movements which take place in individual leaves, so as to distinguish them from those aforementioned ones found in whole leaf-sequences. *Figure 1* shows the development of one of the first leaves of *Cardamine hirsuta*, Hairy Bittercress. The process begins with the emergence out of the growing point of a smaller pointed structure, which gradually enlarges and is soon joined by four similar pointed protrusions. Between these the leaf-blade then fills out, encompassing the whole periphery. Meanwhile the base of the leaf has been narrowing down, "bunching itself up", and the shoot-like structure thus formed grows then in such a way that the blade is lifted outwards. The points are still protruding somewhat, but they gradually become rounder and plumper, before being finally "rounded off". The previously notched blade then appears uniformly round, and by this time has distinguished itself clearly from the stalk.

The processes involved in the growth of a leaf can be described in various ways. Often a distinction is drawn between lengthening, spreading and thickening. Behind this lies the more or less clear intention to apply the system of Cartesian co-ordinates to the developing leaf. Distinctions can also be made according to

* Translator's Note: The term actually used by Splechtner was *Bildebewegungen* and in previous translations these have been called "formative movements". In such a context, however, "formative" is not concrete enough, whereas "morphic", meaning "morphising" or "morphological-structure-creating", seemed very apt. It also has the virtues of freshness, simplicity and contemporary relevance, being used, for instance, by Rupert Sheldrake in a very similar way.

the location of the growth zone, e.g. between intercalary growth and that at the leaf-tip. *Here* the attempt will be made to single out from the total process various formative tendencies which growth follows.

In speaking of point, stalk and blade, we have a picture of circumscribed form-elements. If, instead of this, we concentrate on the transformatory process going on between them, they then appear as something which we are inclined to think of as formative tendencies, but would do better to describe as formative *activities*. Coming to grips with these activities requires inner effort, for it is only by following through these formative changes through in our imagination that they become truly observable as movements.

The first of the activities thus observed may be termed *shooting*. A pointed structure emerges and grows in a distinct direction. As soon as this event begins to multiply, *dividing** can be said to have set in. For the filling out of the leaf-blade the word *spreading* (because it is derived from the old German word for this) seems appropriate**. The activity by which the base of the leaf is first lengthened and then gathers itself together, while growing shoot-like, to form a stalk, may be called *elongation*.

The process of leaf-formation thus reveals itself to us as a temporally and spatially differentiated interplay of four formative activities. In how far it is justifiable, and indeed essential, to distinguish these four activities will become gradually clearer in what follows.

It has been strongly emphasised in the past (e.g. Goebel 1928, Troll 1939) that both the lowest and highest leaves should be seen as "inhibited" in some way in comparison to the middle leaves in a normal sequence, these latter being the ones in which the archetype is most fully realised. While broadly agreeing with this one can still arrive at the question as to the nature of these "inhibitions". Could it not be the case that precisely through the way they affect various organs regularities are revealed, by which the dynamics involved in the life a plant can be better understood? Do not inhibitions imply corresponding "stimulations"?

Answering this question affirmatively will depend on whether there are grounds for regarding the upper and lower leaves as special expressions of the archetype, and not simply as incomplete. Such a view would make it easier to discern, through the various steps in the formative process, the archetype as a composite, but flexible whole, rather than to fix it to one particular form.

If we observe the morphic movements in the way described, that is, by following the interplay of formative activities, we will notice, even in the individual

---

* Translator's Footnote: Dividing – this has been previously translated as "indentation"; however, the latter rendering does not square with the description of the process as given later in this paper. "Indentation" is more the consequence of the dividing process. "Dividing" has also been preferred to "division" for obvious reasons. The German word involved here is *Gliedern*, and the normal botanic term in English for a leaf which is *"gegliedert"* is "divided". This happy combination of accuracy and botanic custom was considered irresistible.

** The German word here is *spreiten*, which ties in very well with the word for blade, *Blattspreite*, first used by K.F. Schimper (Troll 1939).

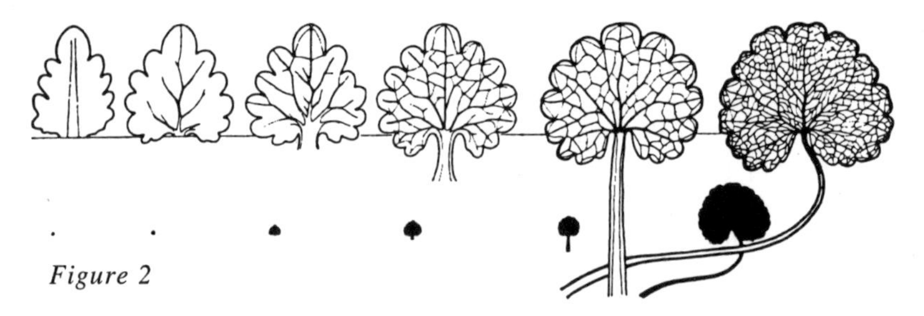

Figure 2

leaf, that certain activities set in earlier, others later, and that they also fade out at different times. The form of the leaf at each stage is related to the duration and intensity of these various activities.

In *Cardamine hirsuta* we saw at first points arise through shooting and dividing; while spreading is engaged in filling out the leaf-blade, shooting fades out. The points persist for a while as little edge protrusions, and then disappear almost totally into the blade. Such interaction between shooting and spreading can vary considerably in the course it takes, and is thus a strong determining factor in the eventual shape of the leaf.

In *Figure 2* the growth pattern of one of the basal leaves of *Glechoma hederacea* (Ground Ivy) is displayed. To facilitate comparison the relative sizes of the series of forms have been altered. This produces certain distortions of the natural forms. For our purposes it was found most convenient to scale everything up to the size of the spreading stage. For comparison, the true sizes are given in black underneath. The first of the stages illustrated shows a number of points which are still growing (magnification makes them appear rounded), while spreading has already begun. Shooting and dividing continue to work upon the lower edge as far as the fourth stage, and then largely come to rest. The rounded bulges of the leaf-edge at this fourth stage are in the grip of the spreading process, which has been growing steadily in intensity. As if tied with strings to the apex of each bulge and the bases of the notches between them the blade "swells up", leaving behind parallel veins as it does so, which appear like ripples. The outcome of this is that the places that formerly were points are now indentations.

The interactions thus described also characterise the manner in which the leaf as a whole changes shape. Beginning ovate, it gradually forms itself into a broad, kidney-shaped blade, and a long stalk. The finished leaf displays almost exclusively the features of spreading and elongation.

2 *Morphic Movements of Leaf-Sequences*

If we now bring the method we have been using to bear upon a sequence of finished leaf-forms, it is immediately clear from the previous examples that this will yield a completely different picture of a morphic movement than that obtained from studying an individual leaf. Troll (1939, p. 962) refers to Wretschko's remark, in a study of Umbelliferous leaves, to the effect that "the leaves up to the umbel represent in turn, as it were, the separate developmental stages of a single basal leaf". Other plants yield similar observations, and this can naturally be seen as supporting the idea, already mentioned, that the upper leaves are inhibited in comparison to those lower down. There are, however, other ways of looking at this phenomenon.

If we consider once more the formative sequence found in *Glechoma* (Fig. 2), we get an approximate picture of the plant's whole sequence of leaves right up into the blossom, but, significantly, in the opposite order. This fact that *the morphic movements of the individual leaf and of the vegetative leaves as a whole run in opposite directions* will concern us further. It should also be observed that the natural leaf-sequence resembles the enlarged forms of Figure 2 more closely than their life-sized, black counterparts. For its own purposes, it would seem, it carries out its own "distortion" of the natural forms.

The arrangement of leaves shown in *Figure 3*, in which the developmental series begins at the bottom left with the cotyledon and finishes bottom right with the highest leaf prior to the blossom, has been used previously (Bockemühl 1982) to bring out the salient qualities of morphic movements. The diagram has been arranged in a loop or lemniscate (not a half-circle!) to indicate how the movement is part of a more comprehensive one, which, emerging from the fruit and the seed

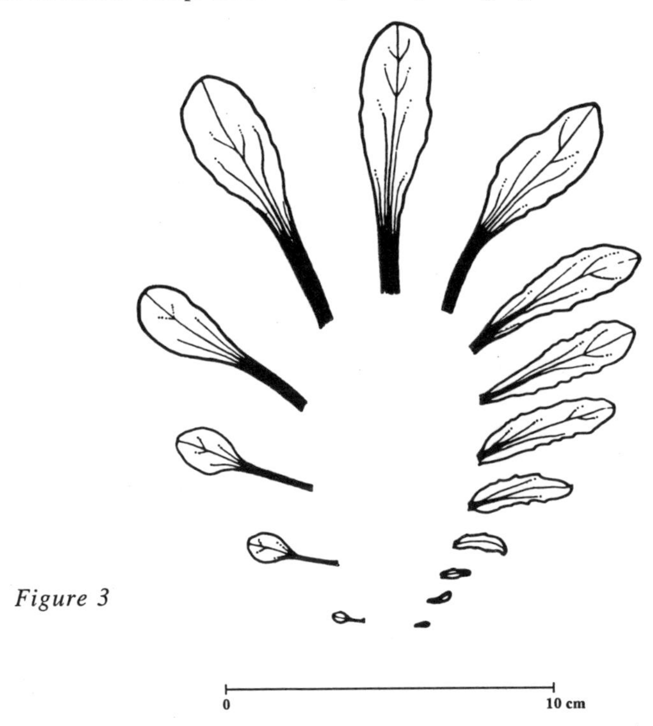

Figure 3

Figure 4

and culminating again in the blossom, goes through a sort of zero point, or "involution"* (on this see the essays by A. Suchantke and R. Buensow). The sequence of forms found in *Valerianella locusta*, the Cornsalad (Fig. 3) is very clear and simple. At the outset stalk and blade are clearly distinguished from each other. These two separate form-elements then merge more and more towards the end of the developmental process. In between, this process of interpenetration can be seen going through several stages: step by step the blade advances towards the base. In doing so it elongates, while at the same time, as it were, absorbing the stalk.

Here also we start with form elements perceived through the senses, and can then transform them in imagination into an interplay of various formative activities. In this context we are even less tempted than with the other examples to identify these activities with growth processes. It is much more readily apparent that the morphic movement now under consideration takes place on an ideal level.

Of the by now familiar activities, elongation and spreading appear most strongly in *Valerianella*. Indeed it provides a very simple example whereby we can plainly follow the interaction of these two activities. In doing this, we also notice that they in turn are subject to higher *regulative principles*. Goethe, when he spoke of "expansion" and "contraction", was the first to identify these principles

* The German word here is *Umstülpung*, for which "involution" is a bit vague, but short of inventing something like "introvolution" nothing better suggested itself.

fully. During expansion, elongation and spreading are more or less *separate*, while in the contraction phase they are *merged*. This can be regarded as a general rule. Even if it should happen that at the contraction phase a part of the stalk remains distinct, the blade will be found to show a definite tendency to lengthen, i.e. to incorporate the elongation process.

In the middle section of the plant the two activities *interpenetrate*, and in each species they will do this in a different way, with corresponding effects upon its appearance. Each species thus displays its own special *motif*, which the morphic movement modifies in various ways. These, it will be found on looking more closely at Figure 2, cannot be fully accounted for, as was also the case with the individual leaf, by elongation and spreading alone. In the contraction phase the hint of a rippled edge and slightly more pointed shape show clearly that dividing and shooting are also involved.

In the finished leaf of *Cardamine hirsuta* we had an example of how it can come about that shooting and dividing leave no visible trace of their activity. Figure 1, however, shows only the development of a lower leaf. As we can readily see from *Figure 4*, things are rather different when we look at a whole sequence of leaves. Here too, elongation and spreading are predominant at the start. Almost immediately, however, the dividing process begins, and in such a thorough way that the elongation activity is able to invade each individual element thus formed. These basic elements then multiply. Elongation and spreading interpenetrate in

Figure 5

such a way that they seem to remain separate. It is as if the whole leaf-blade, in spreading, has been taken over from the centre outwards by the elongation activity. Spreading is only visible on the stalked, pinnate leaflets thus formed. These leaflets, at first round, gradually become divided, while later lengthening and growing more pointed. Thus dividing is followed by shooting. The four activities — *elongation, spreading, dividing, shooting* — thus make their appearance in the opposite order to that found in the morphic movement of a single lower leaf. In the course of expansion and contraction there are other *regulative principles* which they also follow. In relation to the first two activities (elongation and spreading) we have already become familiar with *separation, interpenetration* and *merging*. The two others, dividing and especially shooting, can, as we saw, remain hidden during the expansion phase. They are, as it were, masked by the spreading process. During the contraction phase, however, as elongation and spreading merge more and more into a single activity thus lengthening the leaf as a whole, the leaflets upon which dividing has been working become subject to shooting, and this also has its effect upon the eventual shape of the whole leaf.

In this manner the last leaf in the sequence under consideration turns out to be endowed with form elements that render it the polar opposite of the first one. A good example of such a polarity is found in the leaf-sequence of a single year's growth of *Medicago sativa*, Lucerne (*Figure 5*). Here a threefold structure persists through the whole sequence. Only the uppermost leaf is uniform. All structures, both the leaflets and the bracts, are initially rounded, plump, and later are elongated and pointed. The most astonishing thing, however, is that what starts out at the apex of each leaflet as a notch ends up as a point. This phenomenon is immediately reminiscent of what occurs in *Glechoma* (Fig. 2). Here the notches on the finished leaf are the remnants of early dividing, while the points that initially lay between them are overtaken by spreading to such an extent that they end up as little indentations. The notches on the first leaves of *Medicago* correspond to such indentations, having arisen in the same way, whereas in the contraction phase it is shooting which finally prevails. Once again we observe that the morphic movements run in opposite directions.

With this the question reappears as to the extent to which these four activities can be separated. The difference between shooting and spreading is immediately obvious. Growth in a distinct direction can easily be contrasted with the expansion of a whole surface. It is considerably more difficult to understand the difference between shooting and elongation. Since they both behave in a similar way, one is at first inclined to regard them as one activity, which simply acts upon different areas of the leaf with varying effects. In other respects, however, there are essential differences between them. Shooting is characterised by the growth of pointed structures. It tends to ray out from a centre in one direction. As the point grows, the leaf from which it emerges is left behind. Elongation, by contrast, pushes the leaf outwards by means of intercalary growth (primarily at its base). It tends to "direct" the radii, to bind them together into the stalk, or, when it operates within

Figure 6

the actual blade, to make the veins run in parallel along the length of it. Growth, as far as the leaf-form is concerned, runs in two opposite directions by virtue of these two activities.

Through being able to distinguish between shooting and elongation it also becomes possible to understand the essential difference between the typical lower and upper leaves of dicotyledonous annuals. The expansion phase (and with it the lower leaf) is characterised by elongation and spreading. At this stage shooting and dividing are held back. In contrast to this, shooting dominates the contraction phase, and elongation remains in check. If in this region spreading and dividing should make their presence felt to any notable extent, then their activity will be confined to the base of the leaf.

The leaves typical of perennial plants will not be considered in any detail here. They fit very well into our picture as long as they are not placed at the beginning of a new cycle of growth, but rather at the end of the previous one, which is when they actually did emerge. As such "products of contraction" they are, of course, related to the upper leaves of annuals. The latter, however, lead over into the blossom, whereas the former effect the changeover into a further vegetative phase. The renewed expansion this entails will, if anything, affect the base of the leaf, rather than the tip. Most cotyledons behave in a similar way.

To come now to dividing, the effect it has on the shape of the leaf is not derivable from the combined action of the other three activities, which would produce a completely uniform leaf. Dividing is conceptually very similar to separation. Its nature, however, is to produce repetitive structures, whereas separation, as a higher regulative principle, is responsible for directing qualitatively varying tendencies (elongation and spreading) into different regions.

3 Individual Leaf Development in Relation to the Leaf-Sequence

We must now address ourselves to the relationship between the morphic movements involved in the formation of each leaf of a particular plant and the development of the leaf-sequence as a whole. *Figure 6*, using *Sisymbrium officinale,* Hedge Mustard, as an example, gives us an initial impression of such interaction. It shows four plants, which concurrently germinated and grew at the same site, and were harvested one by one at intervals of about one month. Here, although hardly remarked on before, it stands out clearly that each stage in the series of forms is simultaneously a picture of "expansion" and "contraction". In plants 1 and 2 this is particularly noticeable. In plants 3 and 4, however, there are leaves missing due to wilting, and these can be added on by picturing them, or by borrowing the appropriate ones from the other plants (according to the black arrows).

The essential thing here is not that the leaves within a sequence get larger and smaller, it is rather that we find on the left-hand side the typical long-stalked, relatively rounded forms, and on the right the short-stalked and pointed ones. In the centre are found both the largest and most intricately outlined leaves.

If we follow the path of a single leaf (according to the white arrows) we encounter the most complete picture of morphic counter-movement in the lower leaves. This phenomenon is already familiar to us in what emerged from comparing Figures 1 and 2 with Figures 4 and 5. The white arrow between plants 1 and 2 indicates part of the developmental path of a leaf, showing how its relative position changes from the contraction into the expansion phase. In the later leaves the morphic counter-movement becomes less and less evident. All these observations have led us to investigate the whole field of movements in detail.

We chose *Lapsana communis,* Nipplewort, as the object of our research. We grew a large stand of this plant, removed a few at weekly intervals and pressed the leaves. The smallest ones were stuck straight on to a transparent plastic sheet and covered with paper to dry. The whole process took 13 weeks to complete. In this way *Figure 7* was produced.

If we follow the various series of developmental stages from below, then we gain an impression of the whole process as proceeding like a wave from left to

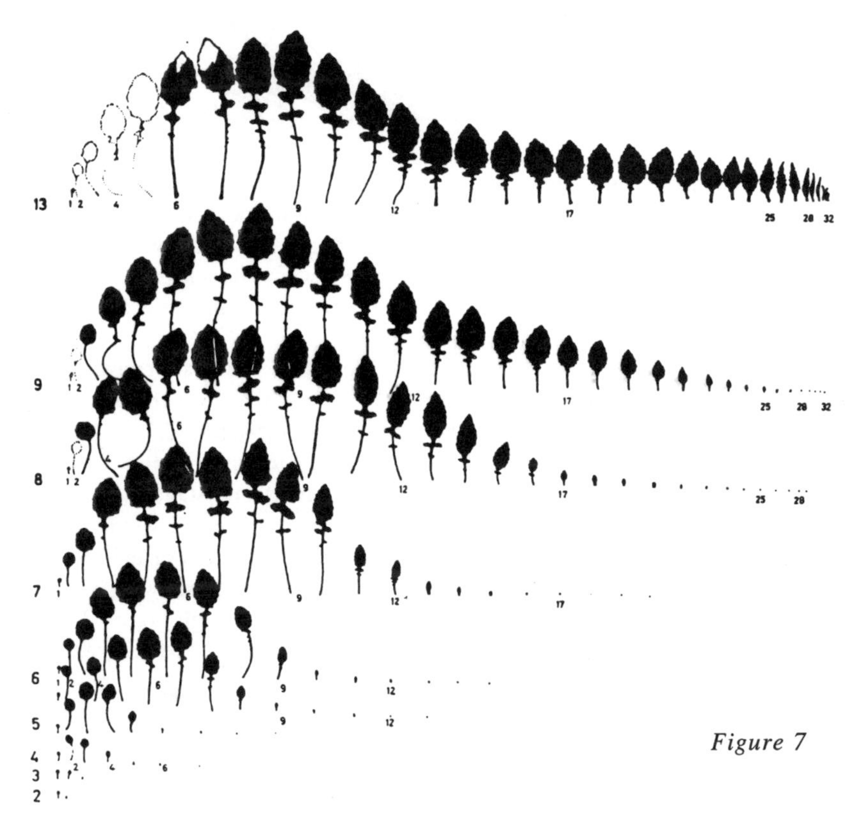

Figure 7

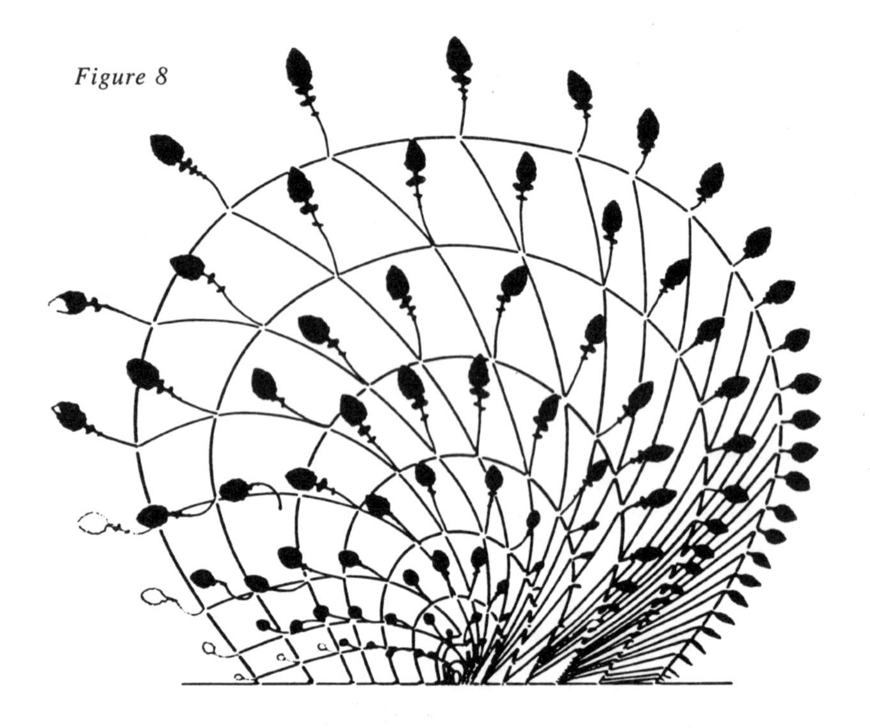

Figure 8

right. This species puts out its lower leaf-nodes very close together, so that the leaves, up to about the twelfth one, are arranged in a rosette. Then the stem shoots upwards and the leaves remain smaller. From this point on the wave of leaf-development is curtailed, and fades out altogether towards the blossom. Before the blossom there is also a wave of branching which cannot be discerned from the diagram.

When these leaf-series are arranged in loops this yields a picture similar to that obtained from *Sisymbrium* (*Figure 8*). As in that case the lower leaves gradually change their position from the right to the left-hand side. This mode of presentation also brings out the fact that the form changes in corresponding fashion.

The process becomes still clearer when the embryonic stages of leaf-development are added to the picture. In *Figure 9* and *Figure 10*, where the same method of illustration has been used as in Figure 2, a selection of individual leaf developments from the leaf-sequences of Figures 7 and 8 is shown. They have been numbered according to the small figures scattered through Figure 7.

The development of each leaf begins with the shooting of a small, fine point (which the strong magnification makes appear rounded). Very early the budding leaf begins spreading laterally. But dividing is faster off the mark, and is responsible for multiplying the shooting points. The beginnings of a stalk also separates out very early on, but elongation sets in in earnest relatively late.

The leaves shown in Figures 9 and 10 fit this general description quite well,

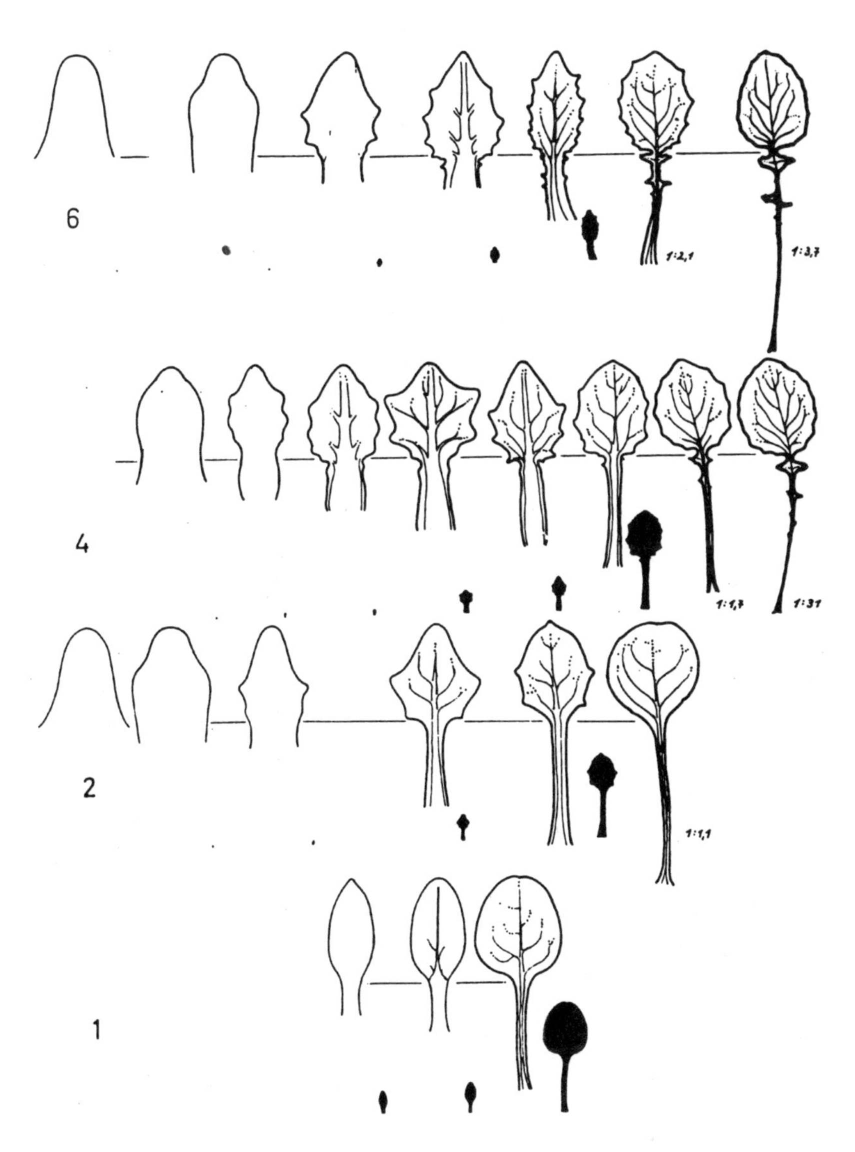

Figure 9

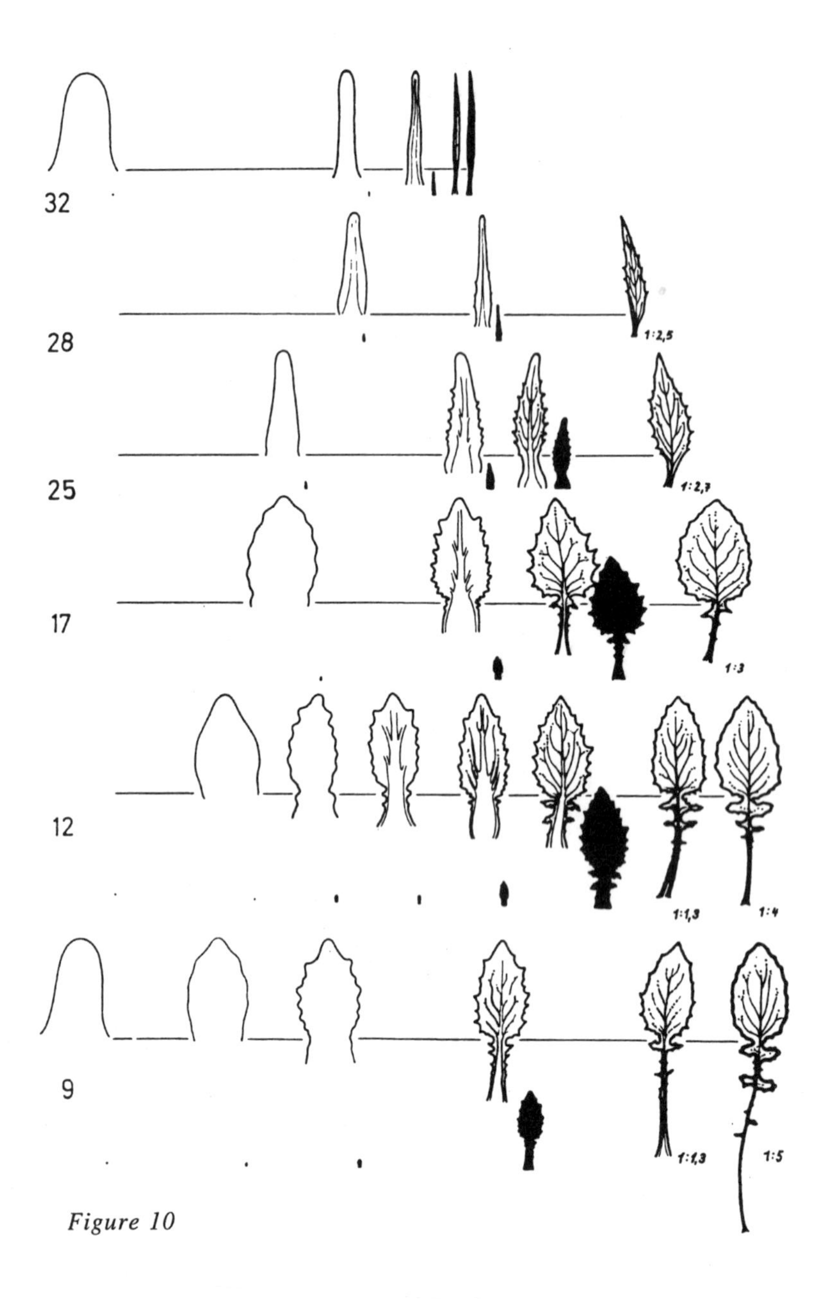

Figure 10

with the possible exception of the cotyledon and the uppermost leaf. The activities vary, however, in their relative intensities. While the shape of leaf 2 is early determined by the appearance of several points, these, as we saw in *Cardamine*, are subsequently smoothed over by the spreading process. This "rounding off" of the points occurs to a much lesser extent in the leaves that follow, so that the effects of dividing persist more and more into the shape of the finished leaf. Moreover, with each succeeding leaf more basal pinnate lobes appear.

In the leaves that grow from the long stem these lobes disappear again; elongation and spreading also become weaker, until shooting is almost the only activity still in evidence. In the region of the uppermost leaf where spreading occurs very slightly, elongation is also taking place. The two processes are merged, and the long, pointed leaf remains undivided. In this it is abundantly clear how shooting, with which the development of every leaf begins, surges gradually to the fore, becoming the dominant activity prior to blossoming. In this connection it is interesting to look more closely at the form of the leaf-tip in relation to that of one of the lateral points as they change from leaf to leaf (*Figure 11*). Each version of the lateral point displays a different relationship between shooting and spreading. It is evident that with every step shooting becomes more pronounced while spreading diminishes.

4 *The Interplay of Morphic Movements*

In *Figure 12* the attempt has been made for the first time to show in one diagram the relationship between the morphic movements just discussed and those of the mature leaf-sequence. It affords a more exact insight into the processes which, in Figures 6 and 8, could only be hinted at. As before, the basic format comes from arranging the mature leaves in a loop. Here all the forms are orientated according to a single loop. From a zero-point, which is at the same time the starting-point for the development of each leaf, radii extend towards the mature forms on the loop. On these radii are arranged developmental stages of different leaves according to their relatedness to the mature forms. When this is done it becomes apparent that each radius joins together a series of forms in which the four activities are present in more or less equal proportions. If plant growth were a linear process, analogous to that of crystals, the actual development of each leaf

Figure 11

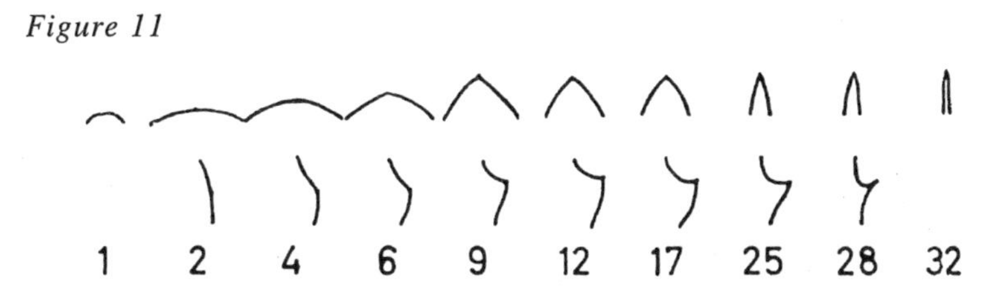

Figure 12

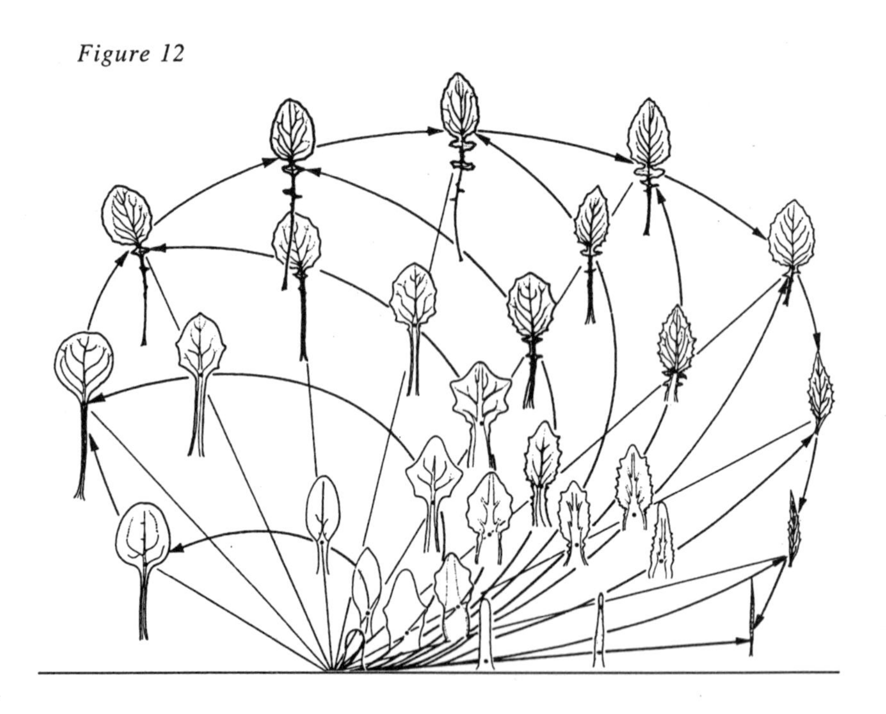

would also follow the radii. In general however, the direction taken by each leaf's development is subject to continual variation.

The first developmental step is the same for all leaves. It serves as a pointer towards the form of the last leaf, which is the one nearest to the blossom (bottom right). This is the only leaf whose complete morphic movement follows this direction almost in a straight line. All other morphic movements curve to a greater or lesser extent. Those belonging to the lower leaves curve the most, and in doing so cross the radii, which connect the later-appearing mature forms straight to the starting point. It could be said, therefore, that the forms at the intersections prefigure these later ones.

The by now familiar leaf-sequence of a mature plant, shown in the outer curve, appears in this light as a series of meeting places for two directionally-opposed movements. Of course, this "motion picture" can only be a pale reflection of the actual processes involved. Nevertheless it gives us an inkling of the archetype as a composite form in movement. In this context the leaf-form typical of the middle part of the plant is only one representation among many, albeit a particularly striking one. This interpretation may indeed prove the key to understanding developmental processes not only in plants, but in other areas as well. In the realm of phylogeny, for instance, could it not facilitate a better understanding of so-called prophetic forms?

It may be briefly mentioned here, that the four formative activities interact with certain environmental factors. The middle leaves of *Lapsana communis* (*Figure 13*) may be taken as an example of how in damp, shaded conditions spreading predominates (left), while through almost complete shading of an older plant (right) elongation can be given the upper hand.

To finish I would like to point out some of the wider implications of the work here described.

Goethe, in pursuing his chosen path in science, was well aware that there were two main ways of straying from it. Either by falling into rigid abstractions, or by getting lost in mystical speculations. We too, in venturing upon the sort of investigation as has been attempted here, must bear this in mind. According to Rudolf Steiner Goethe's work laid the basis for a "rational organicism". He defined the task of this discipline more exactly and in the matter of how advances might be made along the path thus set he gave indications which are still relevant. The observer, he said, must pay attention to his own thinking activity. Doing this reveals what is not accessible to the senses, but which is nevertheless the principle by which sense impressions can be connected.

To be able to approach the true nature of the archetype there is, as it were, a series of observational levels which must be penetrated one by one. As we tried to show (Bockemühl 1982), it is possible, by re-directing attention as described, to get beyond the level of sense impressions, where one is dealing with individual form elements, to a level of sense-free observation where the formative sequence appears as a morphic movement. The thinking process itself performs this morphic movement upon the "leads" provided by the senses. Thus realised inwardly, it is then divided into various activities in the way described. Only now does the process of metamorphosis become concretely available as a tool of thought through which to approach the archetype. Had we remained at the level of form-elements set one beside the other, this

Figure 13

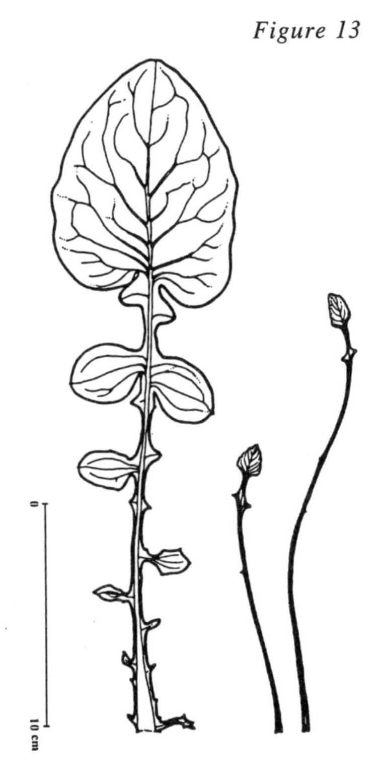

would only have been possible by means of a conceptual model.

The more familiar one becomes with the level of morphic movement, however, the more one realises that here too the archetype has not yet yielded up its full nature. One feels compelled to search for something that imposes order upon the unrestrained behaviour of the various activities. The regulative principles mentioned on page 26 give some impression of this higher level upon which the archetype can reveal itself more concretely. Here too the "motifs" are to be sought, which regulate the specific character of the interpenetration occurring in each species. From the insights we have gained so far, all we can say is that here something is held back, there it is released. In the diversity of movements we sense the presence of a wisdom-filled structure, which presages the concrete reality of the archetype.

3

THE MORPHIC MOVEMENTS OF PLANTS AS EXPRESSIONS OF THE TEMPORAL BODY

Jochen Bockemühl

In the course of its development natural science has devoted much effort to elucidating the properties peculiar to each realm of nature. While it seems obvious that there is an essential difference between, say, inorganic processes and those characteristic of the living world, the constant challenge, nevertheless, remains to describe it in a scientifically satisfactory way.

To manage this adequately individual observations cannot simply be catalogued, but must be integrated in some way into a totality. For instance, it is in the nature of a candle flame to maintain a stable stream of movement, which can be termed a flow-form (*cf.* Hohwald-Haller 1967). By contrast, the morphic movements of a plant are in themselves a constant transformatory process, which follows a higher regulatory principle. That which manifests in space as a stream of movement in the case of the candle-flame, is, in the case of the plant, "disassembled" into a temporal sequence, it has a "time-form".

It might be objected that the events going on within the flame also have a time-dimension. They are, however, all identical; i.e. the process changes in response to external conditions, but not qualitatively from one period of time to another. Time, as it were, only takes hold of such a process outwardly, whereas the plant, in the changes of form it goes through, is not only affected by external conditions, but is itself an expression of inherent periodicity, the components of

Figure 1

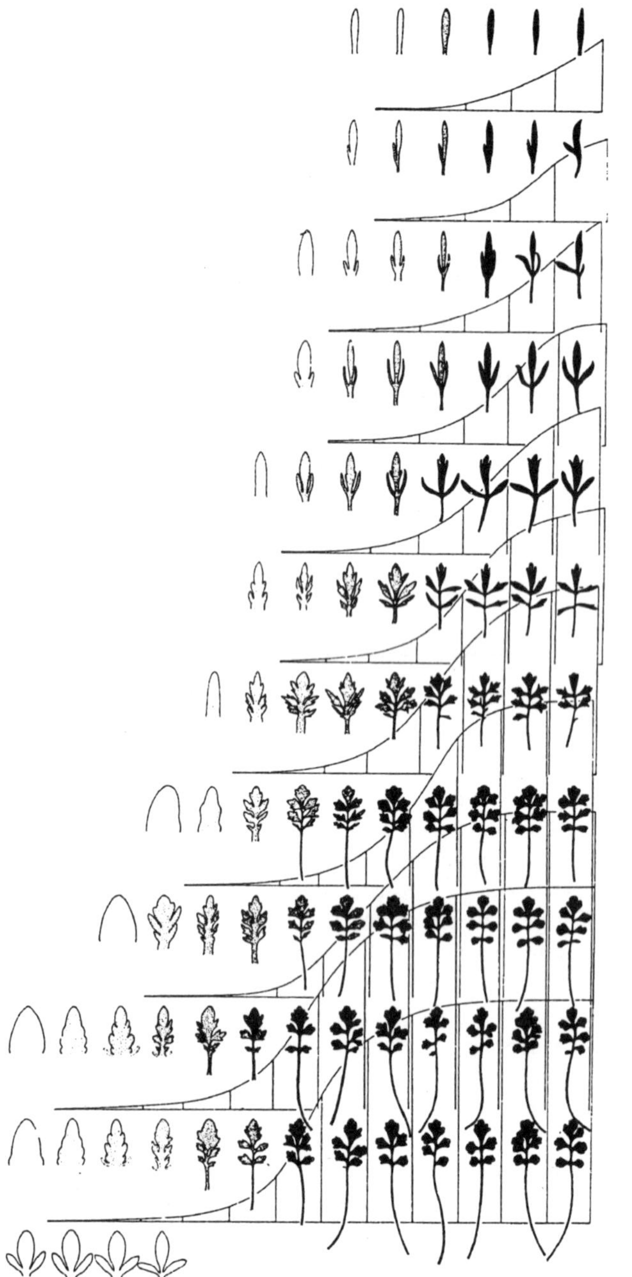

1 2 3 7 9 12 15 18 21 25 30 37 49 Days

which will be the subject of this article.

It is known that in flowering plants the changes in leaf-shape proceed purely vegetatively only as far as that point at which the plant is most "expanded" (Goethe) and the middle leaves are at their most highly differentiated. From the point when the contraction phase sets in the plant is preparing to flower (or to enter a quiescent phase). While it is true that the upper leaves can only form by means of normal growth, in other words the building up of vegetative substance, it is nevertheless apparent that step by step they are held back by the still-unformed flower. The green leaf must pass through a kind of zero-point, before it can appear as blossom (*cf.* Buensow 1982, Grohmann).

The ability to move, thereby effecting a direct change of shape, depends not only upon the accumulation but also upon the breakdown of substance within life-processes. In evolutionary terms this ability first appears in the animal kingdom. Animal movements, however motivated, are ultimately associated with such breakdown processes. They are expressions of an inner realm of soul. Now, in the plant the Goethean concept of contraction can be seen as corresponding to such a breakdown process, and we can observe it in the morphic movement of the leaf-sequence. Only, here it does not become substantial, i.e. it is not a sense-perceptible change of shape, since the normal process of growth, and not the morphic movement, is responsible for the material substance of each new leaf. That which in the animal causes movement and change of form from within, in the plant restricts the growth of the leaves as if from without.

Of course, this observation of regulation "from without" applies not only to the contraction phase, but to the formation of every leaf. Flower formation presupposes not only contraction, but the whole process of vegetative development as well. The fact is simply that with the onset of contraction this restrictive tendency begins to predominate, thereby indicating that flower-formation is imminent. Plants such as ferns, which display only an expansion phase, do not, for that very reason, produce flowers. With them reproduction remains a vegetative process. As such it is detached from the plant, and takes place under damp earth in the prothallium. There are only a few ferns whose sporophylls show a slight contraction, and are for this reason regarded rightly as transitional forms with respect to the flowering plants.

Here already a polarity becomes apparent between two directionally-opposed tendencies underlying plant growth. This can be shown more clearly by placing different morphic movements together for comparison. A previous attempt at this has been made for *Lapsana communis* (Nipplewort, see p. 31). It showed how a form which will only appear later in maturity can be prefigured in one of the immature stages in the morphic movement of an earlier leaf. The latter is also liable to disappear again. The leaf-sequence of a whole plant involves the same succession of forms as that found in the development of an individual leaf, but they unfold in opposite directions.

To elucidate the nature of the interaction between these two movements

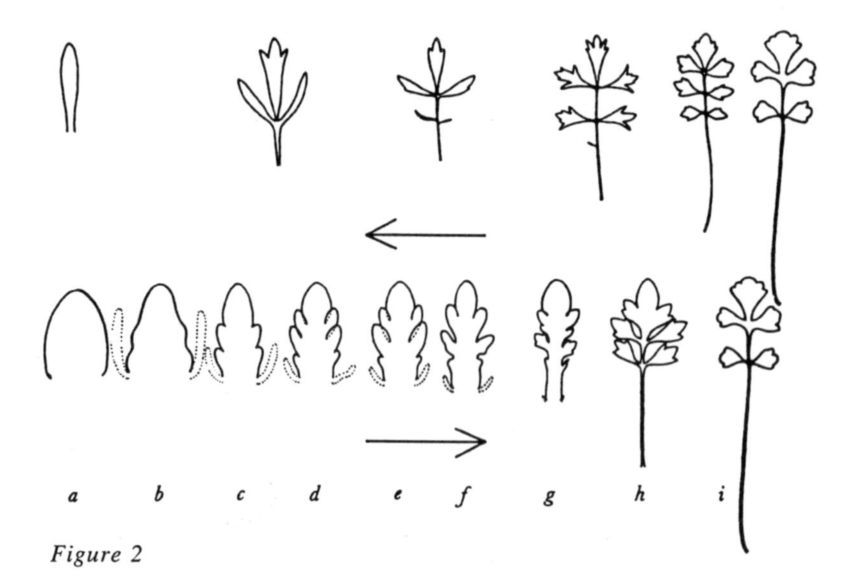

Figure 2

Lepidium sativum (Garden Cress) will now be used as an example. In its own specific way it shows basically the same picture as Nipplewort. The illustration which follows is intended to show the complete developmental stages of all the leaves of one plant. Practically speaking, of course, this meant preparing large series of leaves from plants grown under identical conditions. The results are brought together in *Figure 1*. The developmental stages are shown moving successively from left to right. To facilitate comparison the relative sizes of the leaves have been standardised. The true dimensional relationships between them are shown by the curves. The sequence of leaves goes from the bottom upwards.

In *Figure 2* several steps in the leaf-sequence are set against the developmental stages of the lower leaf. On the basis of this comparison the morphic counter-movements can be made apparent by inwardly changing one form into another and observing in imagination the *activities* thereby set in motion to achieve this. It is thus possible to distinguish four such activities, which usually operate simultaneously, but vary, nevertheless, in their relative intensity. In individual leaf development (Fig. 2), then, the order of their predominance is as follows:

1. *Shooting:* A point (a), which under strong magnification looks like a blunt cone, forms, with a linear tendency away from the plant.
2. *Dividing:* The form elements are multiplied, so that now several points form and splay outwards (b – g).
3. *Spreading:* Here growth takes hold of the leaf-surface as a whole, which thus expands peripherally (f – i). In contrast to Nipplewort, even in the

first leaf of Garden Cress spreading sets in relatively late, so that by the time it does dividing has already been at work inducing the formation of separate leaflets. To the latter, therefore, spreading confines its influence. As the leaf approaches maturity the combined influence of dividing and shooting also retreats, so that the later leaflets appear more rounded than their forerunners.

4 *Elongation:* The main veins of the whole leaf bunch together and lengthen. This is most clearly apparent in the formation of the stalk. Unlike shooting, growth is oriented towards the plant. Since elongation begins here relatively early (f), it participates in the separating of the leaflets. This is the reason why at (i) the lowest pair of leaflets is so far down the stalk.

The sequence of mature leaves shown above in Figure 2 and running from right to left offers a different picture. The effects of elongation and spreading predominate at the outset. Dividing increases in the following stages, but diminishes again, until shooting finally wins the upper hand.

Figure 3 presents a summary of the morphic movements of all the leaves. The outer arc represents from left to right the morphic movement of the mature leaf-sequence. The inner spirals show what takes place in the individual leaves. They all ray out from the same embryonic point a little left of centre, and intersect in their various courses with a series of radii, arranged so as to link those forms in which the activities are seen to be interacting in the same way. Each point at which the outer arc meets one of the inner spirals indicates the position of a mature leaf. Each leaf arises out of the mutual interaction of the two movements. The distance a mature leaf stands along the outer arc is inversely proportional to the tightness of the spiral. By the time the arc has reached the later stages of the whole plant the spiral has become almost straight, and the leaf in which it culminates goes through correspondingly fewer changes of form.

This comprehensive diagram also yields an approximate impression of the plant at any given developmental stage. The dashed spiral, for example, joins together the leaf-forms belonging to a particular moment in the plant's growth. It is evident that in each one of these the plant displays formative features simultaneously characteristic of both expansion and contraction. As development continues this picture becomes increasingly clear, and encompasses a larger number of the plant's organs.

What do these observations tell us about the nature of time?

With our senses we can only perceive time as "one thing after another". In any physical process time moves only in this one direction. In its morphic movements, however, the living organism bears witness to counter-movements in time, which in its process of development are held in balance.

This substantiates a maxim of ancient wisdom, to which Rudolf Steiner often drew attention; namely, that the nature of time cannot be fully grasped unless time as a succession of events in space is complemented by a counter-moving time-

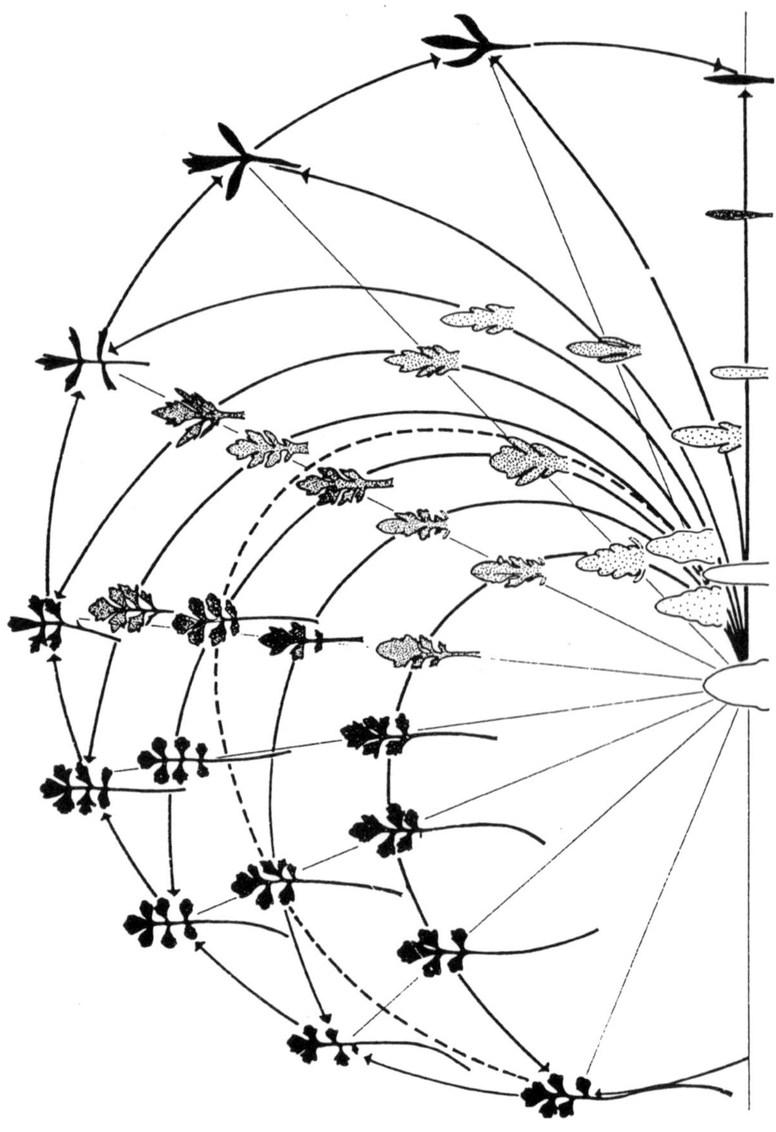

Figure 3

stream, which is only perceptible by means of the imagination. When we mobilise our will by imagining something that needs to be done, we gain a palpable experience of how we are continually inducing this stream into our lives from the future. The closer the deed comes to being accomplished the closer will the mental picture of it come to being transformed into an external "image". By contrast, for the senses the carrying out of the action unfolds in the opposite direction. From another point of view the directions of these two movements can be switched, but they remain as opposed as before.

In the plant's sequence of leaf-forms we encounter a series of ready-made images, which we have taken as a reliable indication of the presence of an *ideal* movement underlying plant structure. The morphic tendencies of this ideal movement appear then in physically continuous form in the developing single leaf in the opposite order. This "material movement", in turn, is put through a series of variations from leaf to leaf. The plant's "youthful urges" * being thus "pictured" in the whole leaf-sequence create an analogy of a mental ability which only appears later, on a higher level of evolution, in animals and man.

From the point of view of inner observation structures have arisen out of qualitatively different activities and taken on the outward features of, on the one hand, the spatial aggregate and, on the other, the temporal succession. If something partially separates itself from its environment in this fashion, then we are justified in using the term body. A body's form and structure is always determined from a level of being higher than itself. The concept, however, must be further differentiated, for we have encountered such discrete entities on several levels. Such differentiated concepts are to be found in Rudolf Steiner's spiritual science, and it seems to us that we are justified in using them in natural science, as long as we can point to their equivalents in nature. Thus the flame presents itself to us as a spatial (or physical) *body*. Changes in it only occur through outward conditions, time remains "external". In the variations continually being wrought upon the plant's spatial body we can discern the expressions of a *temporal body*. The latter appear pictorially in the morphic movements. In spiritual science the temporal body is also designated as the life or etheric body (*cf.* Poppelbaum 1952). It is composed of actively interpenetrating *formative forces*, which we have encountered upon our path of observation as activities. Its structure in turn derives from the higher astral realm, at which we have only hinted here. In the plant, unlike the animal, this does not find any level of bodily expression. It is hoped that it will have become clear from the foregoing that a science which would approach concrete realities demands a faculty of observation which is at home on several levels and requires constant practise.

In this connection reference may be made to the essay on "Biological Thinking" by Wolfgang Schad (1982). In it causal and final explanations are charac-

---

* Translator's note: This somewhat unusual phrase has been used in an attempt to capture something of the pun in the German text. The word used is *Pflanzen-Triebe*, where *Triebe* means both "embryonic leaves" (shoots), and "instincts".

terised according to their epistemological value and assigned to their appropriate areas of relevance. In between these two, assigned to life and the present, is what he called *correlation*. In this paper we have presented an example of how the paradox, expounded by Schad, of the interpenetration of the causal and the final reveals itself in plant development, and how the plant, therefore, must be understood as representing, at each stage in its complex of forms, a correlative whole.

4

THE METAMORPHOSIS OF PLANTS AS AN EXPRESSION OF JUVENILISATION IN THE PROCESS OF EVOLUTION

Andreas Suchantke

The Discovery of Morphic Counter-movements

Goethe's discovery of the metamorphosis of plants has, until very recently, provoked no new research of any depth. For a century and a half little heed was paid to it, except by the morphologist William Troll, whose monumental work[1] has been shunned by the scientific community. This is understandable, for in terms of the nominalistic and reductionist attitude of modern biology the idea of the "archetypal plant" is a cognitive construct, a generalised abstraction, anything, indeed, but a representation of an ideal reality underlying and conditioning material phenomena. It is therefore regarded as old hat. The view is that "causes" reside ultimately in purely material events at the molecular level, and the "totality" of an organism is nothing but an aggregate or "network" of mechanically-caused processes (Heberer 1974)[2].

That there can be no question of a mere aggregate here, is shown by even a cursory glance at the changes that occur in the form of a plant's leaves in the course of its growth (and the same goes for all formative processes). In its shape the individual leaf is no isolated phenomenon, with no relation to the leaves that come before and after it. It rather appears as an integral part of a continuum, which, although varying in form, bears witness to the presence of a larger whole. Actually, to speak of the "individual leaf" is misleading, a product of the human understanding's separative tendency, for ultimately it can only be understood as a section or phase of a temporal process integrated into a clearly-defined whole. In its capacity as a part it is comparable to a single note or chord in a melody. The latter is not a product of cursorily adding one note to another, but rather is itself the higher reality determining the position of the individual note.

A further objection raised against "metamorphosis" is that, if it is supposed to be the perceptible activity of the archetype, why is it only observable in relatively few plants? The archetype, the idea of the plant as such, really ought to be present in all plants, and therefore susceptible, by virtue of their metamorphoses, to "perceptual judgement" *. This is indeed the case, albeit with widely varying intensity. If one takes the trouble to investigate plants which seem at first glance to have uniform leaves, then it will be observed that the familiar variations of leaf-shape do occur, but more discretely, tentatively than in the species of Crowfoot, Scabious, etc., where they stand out very clearly. This goes even for the inconspicuous grasses! [3]

For some years now research into the metamorphosis of plants had enjoyed a new lease of life through a discovery of Jochen Bockemühl's. The material presented here also builds upon this. Bockemühl discovered that the main thrust of the leaf-transformation process — which does not follow any old order, but shows a characteristic overall structure — occurs not only in one direction, but can also reverse itself. This might at first sound rather astonishing. As is well known, biological processes differ from inorganic ones precisely in that they are irreversible. A mature individual cannot change back into its own juvenile or embryonic stage. We now know, however, that this statement — at least when formulated in such simplified terms — must be revised.

Contrary as the two movements are, their opposition is nevertheless indirect, for their respective levels of activity within the plant are different. The already familiar one is seen in *the succession of leaf-formations*. Each leaf as it forms, with its more or less altered shape, represents a step in a total formative stream. As a rule, the first-formed, lower leaves display the roundest blades, or at least come closer to being round than all subsequent leaves; they also have the longest stalks (relative to their total size). The leaves which follow have ever broader and shorter stalks, and undergo successive lateral indentations or dissections resulting in various lobes — leaflet-like, finger-shaped, pointed. In those cases where this does not occur, the whole blade will become narrower and more pointed, the leaf as a whole thus displaying, as it were, what appears in other plants as a part of a leaf arrived at through a series of indentations (see *Figure 1*). Finally, the upper leaves closest to the flower are not only particularly small, but also have the simplest

* Translator's note: German – *anschauende Urteilskraft*. The phrase is Goethe's, and this was the very basis of his method. More about this will be found in the *Introduction* (p.4).

Figure 1: Although varying from species to species, the metamorphosis of the vegetative leaves always follows a similar pattern: the first-formed leaves have a long stalk and more or less rounded blade. Moving upwards stalks become shorter, leaf-blades narrower, either as a whole or through division of the blade into segments. *Above:* a delicate, pink-flowering Bindweed from Greece (*Convolvulus tenuissimus*). *Below*, from left to right: a South American, red-flowering member of the Daisy Family (*Emilia sonchifolia*), the Corn Buttercup (*Ranunculus arvensis*, with cotyledons) and a Southern European Crowfoot (*Ranunculus neapolitanus*).

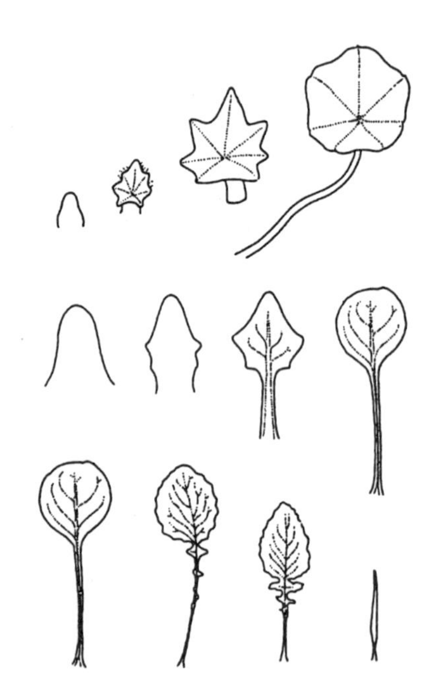

Figure 2: In the upper and middle row from left to right, growth stages of a single leaf within the bud of Nasturtium (*Tropaeolum*, above) and of Nipplewort (*Lapsana communis*); in the bottom row, four mature leaves of Nipplewort beginning with the finished lower leaf from the middle row, and ending with an upper leaf close to the flower. (After Chodat 1920, upper row, and Bockemühl 1966, altered.)

shape: with their broad base attached directly to the stem they simply taper to a point.

The reversal of this process is not found in the sequential formation of different leaves, but in *the development of the individual leaf.* If we take the sequence of events by which the basal, long-stalked, round-shaped leaves form within the bud, then at the first stage, called *"shooting"* (Bockemühl), out of a broad base a little pointed growth, still undifferentiated, arises (see *Figure 2*); after this begins the development of the blade, in that its smooth edge becomes relatively uneven, at least more so than will be the case at a later stage. The various indentations resulting from this "*dividing*" process fill out again in the next stage rendering the blade more or less rounded in shape (*"spreading"*). Lastly, through the process of *"elongation"* the stalk lengthens, lifting the finished leaf outwards.

The main distinction between the two sequences of events just described is not that they run in opposite directions. They differ also both in the level at which they act, and in character. What occurs in the succession of leaf formations along the whole plant is *physically discontinuous:* the finished leaf does not change into the following one, but rather this latter takes up where the previous one left off and carries the formative process further; the previous leaf remains unchanged. By contrast, inside the bud *physical continuity* reigns in the development of the individual leaf, which proceeds smoothly, one and the same organ going through a continuous stream of change. At the level of the individual leaf, therefore, the development acts physically within the material structure of the plant. In the former case, however, the process of change takes place in that sphere of formative

Figure 3: Fern species from the Palaeozoic (Devonian and Carboniferous). Above right, *Rhynia* (after Kidston and Lang 1921), below centre, its relative *Hicklingia* (after Zimmermann 1969). Above left, *Asteroxylon elberfeldense* from the Wuppertal area (after Mägdefrau 1968). Below left, *Stauropteris* (after Mägdefrau), right, *Pseudosporochnus* (after Zimmermann). See text.

causation which underlies (or "oversees") the physical structure, and the actual transformation occurs not in the physical organs themselves, but *between them*, where there are *no* physical structures. This realm underlying the visible structures, where the transformation process takes place, consists of a configuration of activities which may collectively be called the morphic system. It was said at the outset that we are dealing here not with an aggregate of events following fortuitously one upon the other, but with a series of structures in time, each one an integral part, and manifestation, of a larger whole. This being the case, we are justified, also at this level of the plant, in speaking of a "body" (in the sense of individualised form) — of a morphic body, or temporal body (or etheric body).

If these two directionally-opposed morphic movements are compared — the formation of the whole plant in the ordered sequence of different leaves, on the one hand, with the moulding of the individual budding leaf on the other — then it can be said that *in the course of its ripening, the plant, in growing older, arrives at ever more juvenile stages.* The lowest leaves, formed earliest in the plant's development, end up as the ripest, "oldest", while the middle, characteristically-shaped ones finish at a more juvenile stage, and finally the latest-formed leaves stop at the most juvenile stage of all. Paradoxical as it may seem, then, the plant gets younger to the extent that it ages! Its physical maturing is correlated to an increasing tendency to arrest organ formation at ever more immature stages. In other words, the more the plant ages physically, the more it holds in reserve its formative forces, which then are less and less involved in the moulding of physical structures, and are correspondingly able to preserve their formative potential in its pristine state. The morphic body becomes young while the physical body is ageing.

The Development of the Leaf through the Ages of the Earth

Up to this point formative process has been considered on two levels: that of the development of the individual organ, the single leaf, in other words, *organogenesis*, and that of the unfolding of the whole organism, or *ontogenesis*. Out of this the question quite naturally arises as to how these two morphic countermovements relate to the third, still higher level of *phylogenesis*, i.e. to the evolution of the plant kingdom as a whole. In terms of the biogenetic law — according to which ontogenesis is an abridged, sketchy recapitulation of phylogenesis — there should be traceable parallels between what happens at the level of the single plant and the large-scale development of the various phyla. Assuming recapitulation occurs, however, the question remains as to which direction it follows.

To give an immediate answer, the apparently reasonable assumption that it must follow the "metamorphosis", since the succession of differently-shaped leaves is, after all, the plant's ontogenesis, is incorrect. Rather, it is the course of

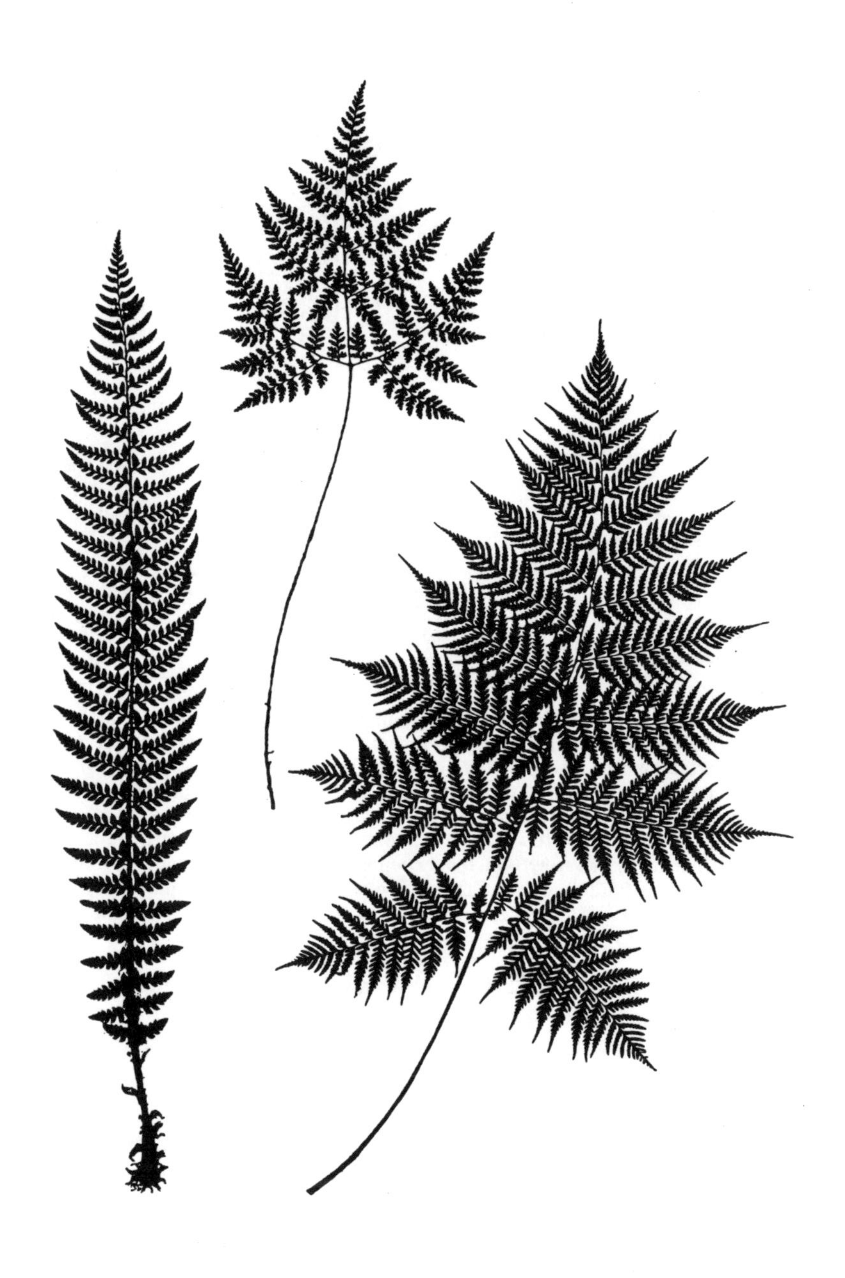

Figure 4: The pinnate fronds of ferns are not simply repetitive structures, but composite leaf-forms, in which individual elements are subordinated to a formative whole. Three living species: left, hard Shield Fern (*Polystichum aculeatum*), above, Mountain Bladder Fern (*Cystopteris montana*), right, Canary Lady Fern (*Diplazium caudatum*).

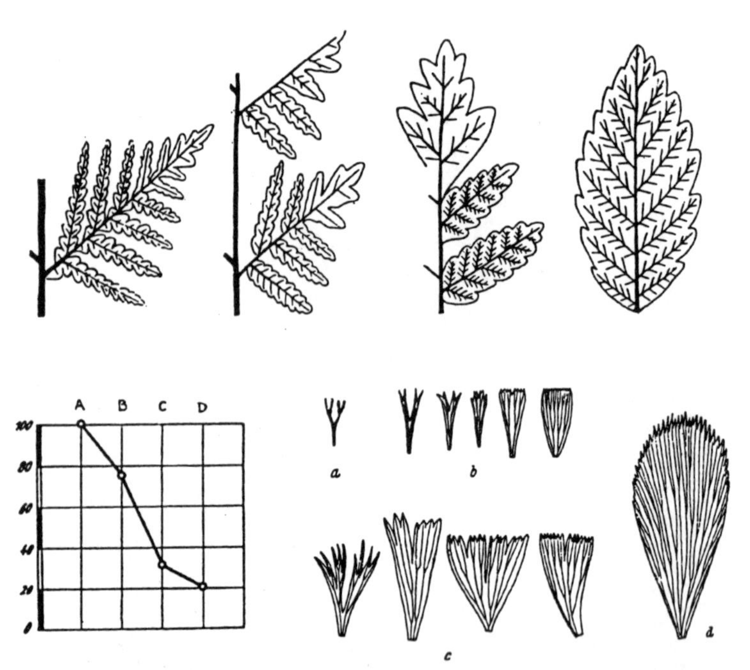

Figure 5: Above: From dividing to spreading. Frond development in the fern family of the *Emplectopteridae* from successive layers of the Permian period (after Asama 1962). Below: Curtailment of free-growing shoots and transition to the closed leaf-blade in the fern group *Sphenophyllum* as shown through various geological periods. Horizontal axis in the graph: A: Upper Devonian, B: Lower Carboniferous, C: Upper Carboniferous, D: Rotliegende (Permian); vertical axis: percentages of known species with free-growing shoots. Right: a. *Sphenophyllum tenerrimum* (Lower to early Upper Carboniferous), b. *Sph. cuneifolium* (middle Upper Carboniferous), c. *Sph. maius* (middle Upper Carboniferous), d. *Sph. thoni* (Upper Carboniferous to Rotliegende. (After Mägdefrau 1968 and Zimmermann 1969.)

organogenesis, i.e. the physically continuous development of the individual organ, which corresponds to phylogenesis. *The metamorphosis must therefore be seen as reversed phylogenesis.* And this being the case, it must be possible to speak of a *reversed biogenetic law!* We will deal with these facts in more detail later on.

To provide a background to the question under consideration what follows is a (necessarily brief) survey of the development of the leaf in the course of the Earth's history.

The earliest, more or less amphibious "land plants" were strangers to the difference between shoot and leaf. For instance, the archaic fern *Rhynia* from the Devonian period (*Figure 3*) had identical, forked, upright shoots, some of which carried spores, while others remained purely vegetative. Plants at this stage of development could not yet be said to have leaves, "rather the whole plant consisted of nothing but shoots, all identical" (Zimmermann)[4]. This state of affairs persisted

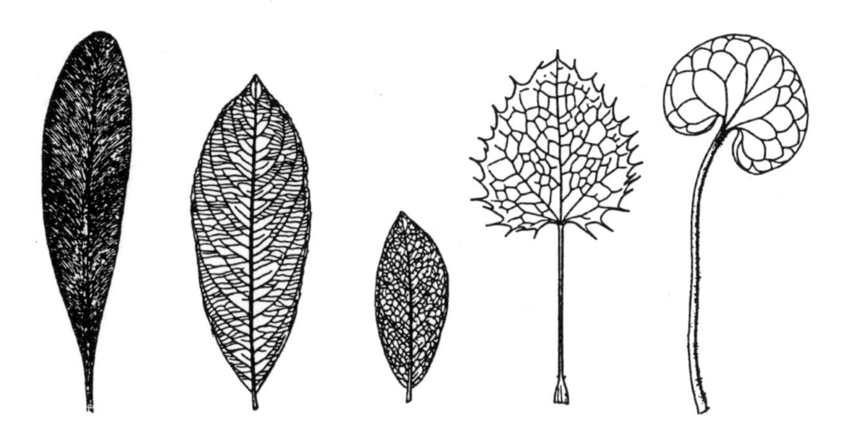

Figure 6: Netted leaves of varying intricacy. From left: *Glossopteris*, a Pteridosperm (Seed Fern) from the Permian. The others are all living forms: Willow sp. *Salix cinerea* and *S. myrtilloides,* Barbary, *Berberis vulgaris*, Asarabacca, *Asarum europaeum.*

through the transition to apparently more complex structures, as seen in two other Devonian forms, *Asteroxylon* and *Pseudosporochnus* (Fig. 3). All structures on these plants stayed purely at the level of "shooting", which was multiplied, as it were, endlessly. *Asteroxylon*, and to a greater extent, *Pseudosporochnus* showed an increasing centralisation in their pattern of growth. The tendency towards uprightness typical of the land plant became more pronounced, but off-shoots, simply carrying on the same forked growth-pattern sidewards rather than upwards, showed no sign of becoming differentiated leaf-structures — "the difference between vertical and lateral organs is negligible" (Zimmermann). The same goes even for the primitive fern *Stauropteris* from the Westphalian Carboniferous (Fig. 3). What at first glance looks like a normal frond is actually a group of perpetually-branching shoots, each new branch going off at right-angles to the one before it. This results in a bushy structure, which is not in the least like the two-dimensional, pinnate frond familiar in the higher ferns (*Figure 4*). We have still a long way to go before reaching the developmental stage of the *divided leaf.* The concept of dividing implies, after all, a whole which is differentiated, in other words its parts *vary* in form. Whereas what we see in the primeval, leaf-*like* forms is nothing other than endless repetition of a single structure. "The more imperfect a created organism is, the more its parts resemble both each other and the whole. The more perfect a created organism is, the more unlike each other are its parts. In the former case, the whole is more or less identical to the parts, in the latter parts and whole are dissimilar. The more similar the parts are, the less subordinate are they one to another. Subordination among the parts is an indication of perfection" (Goethe)[5].

Genuine differentiation first appears in the late Palaeozoic era in the fronds of those ferns typical of the Carboniferous and Permian periods, the morphological

Figure 7: Phylogenesis of the Ginkgo leaf. From left to right: *Glossophyllum florini*, Keuper; *Sphenobaiera furcata,* Keuper; *Sphenobaiera digitata*, Zechstein; *Baiera muensteriana*, Rhaetian (Upper Triassis)/Lias; *Baiera brauniana*, Wealden, *Ginkyoites pluripartitus*, Wealden; two leaves from *Ginkgo adiantoides*, Pliocene (after Krausel 1953 and Mägdefrau 1968).

characteristics of which have been preserved in the "modern" ferns. Certainly there is still no question here of a true leaf, as we find it in the higher flowering-plants. What we have is a structure simultaneously bearing features of a shoot (the growth zone is at the tip rather than the base), and reproductive organs in the form of spores. That which the flowering-plants separate out into shoot, flower and leaf, are here found in a state of primal unity. Nevertheless, the frond *does* possess some unmistakable leaf characteristics, such as its composite shape, to which its parts, the primary and secondary leaflets, are subordinated; or again the typical differentiation into an upper and a lower side.

Into this period also falls the transition to the closed leaf-blade, in other words to the spreading stage in the morphogenetic series. The transition occurs either from the pinnate (divided) stage, or directly from the previously described shooting stage (see *Figure 5*). Various levels or degrees of Goethean perfection are discernible therein: in *Sphenophyllum* what we see is basically a joining up of identical elements to form a simple, uniform blade, lacking any further differentiation. In the *Emplectopterida* this is repeated on the level of dividing, such that even in the closed blade the venation pattern of the leaflets is preserved. The

Figure 8: Ontogenesis of the Ginkgo leaf. *Ginkgo biloba*, a living species. From left to right: leaf from a young tree 1m. tall, leaves from a long and from a short shoot of a mature tree.

netted leaf (*Figure 6*), coming next, represents the highest level reached by the spreading stage. Its structure reflects all three developmental steps: shooting in the central axis, the midrib, dividing in the quasi-pinnate side-veins raying out from it, spreading in the formation of the vein network, the anastomoses. Spreading has made of these developmental elements an integrated whole. Elongation, the fourth and last phase, entailing the formation of a distinct stalk, is not found (or at best only rudimentarily) even in those Carboniferous relatives of the ferns, in which the leaf-blade is already fully developed. This is a structure typical of the most highly developed flowering-plants, the Angiosperms, which do not appear until later.

These developmental steps, however, are not confined solely to fern species. They are closely matched by primeval gymnospermic flowering-plants, such as the Ginkgos (*Ginkgoales*), in which they can be followed from the end of the Palaeozoic right up to the sole survivor of today, *Ginkgo biloba* (*Figure 7*).

The examples could be multiplied at will and would always demonstrate the same thing: *the phylogenetic development of the leaf runs counter to the "metamorphosis"* — i.e. to the formative sequence the various leaves of a plant undergo. Now, does this mean that, plant ontogenesis being diametrically opposed to phylogenesis, the biogenetic law has therefore lost its validity for the whole plant kingdom? Inconceivable! It is much more likely that *the metamorphosis cannot simply be taken as the equivalent of ontogenesis and is quite possibly of a totally different nature belonging to a different level.* We need look no further than the Ginkgo tree for the first concrete indication in this connection. In growing from seedling to maturity, in other words during its ontogenesis, it goes through a leaf-sequence, which, in a tentative but clearly recognisable way, recapitulates the phylogenetic development which has taken place in its taxonomic grouping since the late *Palaeozoic* (*Figure 8*).

This might seem to confuse the picture somewhat, but a look at the trees, the woody plants among the Angiosperms, clarifies and resolves the apparent contradiction. If we consider the development of a vegetative shoot in any species of tree or shrub, be it rose or cherry, apple or peony (*Figure 9*), the sequence already familiar from the descriptions of both phylogenesis and organogenesis (individual leaf-development) is quite clearly recognisable. The process begins with the bud-scales producing small, undifferentiated, rudimentary structures. These keep the

Figure 9: Directionally-opposed morphic movements in the sequence of leaves from bud-scale to vegetative leaf (left half of both rows) and from there via the calyx to the petal (right); above: Peony (*Paeonia moutan*); below: Garden Rose (*Rosa*). In both cases the intervening sequence of mature leaves has been omitted.

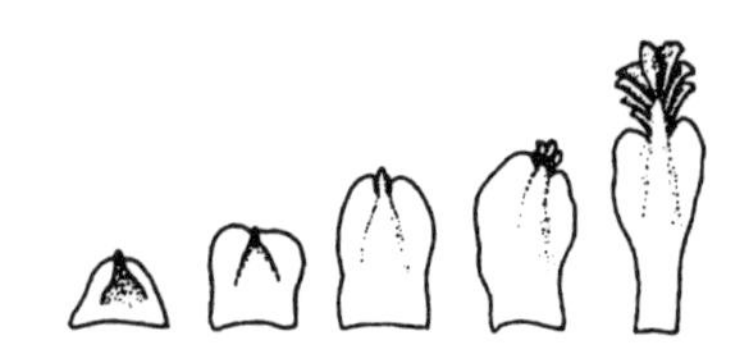

characteristics of shooting until the leaf gradually forms out of the tip and grows upwards, at first small and indistinct, but then larger and more imposing. In the development of the apple leaf (*Figure 10*) it can clearly be seen how shooting is immediately followed by the dividing stage (where, incidentally, the rose leaf finishes). It appears in the two stipules, which are discarded when the blade spreads.

Up to this point we have the same contradictory picture, *but if the shoot culminates in one or more blossoms the process will continue by reversing the sequence just described.* In this we observe not only that the subsequent structures are progressively reduced in size, but more particularly that they are arrested at an ever more immature stage (Suchantke 1982, Schad 1989)[6]. Here we have before us the phenomenon we encountered before, albeit more pronounced and differentiated, in the "metamorphosis" of herbaceous plants. The patent conclusion to be drawn from Figures 9 and 10, then, is that the *proximity of the blossom* is the decisive factor, regardless of the nature of the organs undergoing metamorphosis. These can be bracts situated just below the blossom (Apple), sepals (Rhododendron), calyx (Rose), or those structures which are part bract, part sepal (Tree Peony). *Therefore the influence which reverses the formative sequence of phylogenesis emanates from the blossom;* the "reversed Biogenetic Law" is an expression of the flowering-impulse.

In this connection the difference between tree and herb is astonishing. It is not just that in the case of the tree the "reversal" remains indistinct in comparison to its full expression in herbaceous plants. Much more surprising is the fact that the phylogenetically "correct" formative sequence, considerably more marked in trees than the "reversal", *is completely suppressed in herbaceous plants:* they begin at the highest, most differentiated stage of development, as the first leaves they put out, with their rounded blades and long stalks, clearly show (*Figure 11*). All the preceding stages, which find structural expression in trees, are missing.

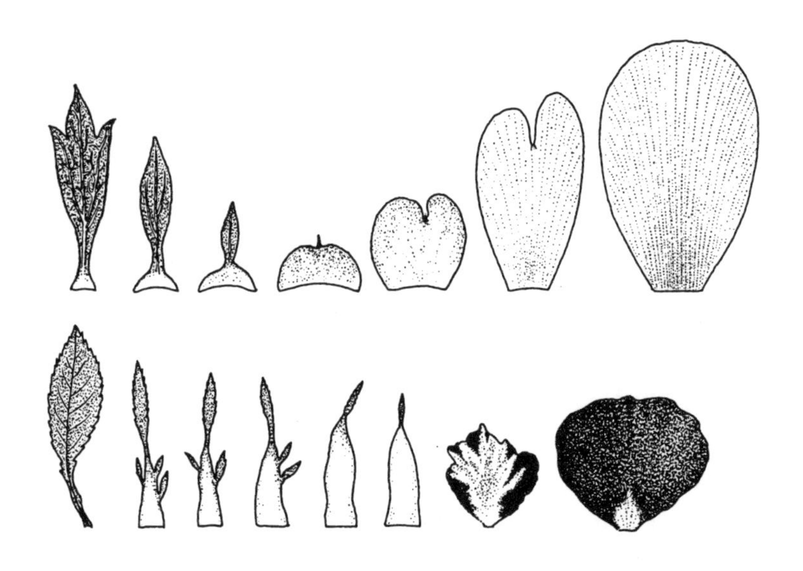

Evolutionary Counter-movements

In order to come to terms with these two phenomena — the reversal of the phylogenetic sequence in connection with flower-development, and the divergence between tree and herb — it is necessary to cast an eye briefly upon two antagonistic tendencies that play a significant part in the evolution of living things.

The first is known as "anagenesis" (evolutionary progress). This process involves initially simple structures undergoing ever-increasing differentiation in

Figure 10: From shooting, via dividing, to spreading and elongation. Leaf-sequence from bud-scale to mature leaf in Apple (*Malus domesticus*). The latter is not to scale and should be much bigger. At the base of its stalk are the two scars left by the divided stipules. In discarding them the dividing stage is superseded (after Troll, amended).

Figure 11: In herbaceous plants metamorphosis only occurs in flowering shoots. Young, non-flowering shoots tend to put out leaves all the same shape as the lowest leaves on the flowering shoots; they thus display the most highly differentiated and phylogenetically recent leaf-form. Left: leaf-sequence of a flowering shoot (above) and a non-flowering rosette (below) of the Scabius sp. *Scabiosa lucida*; right: Asian Crowfoot, *Ranunculus asiaticus*, a young, non-flowering, and a flowering sample.

form and function with a corresponding increase in interlinking and mutual dependence among organs, all under the direction of the morphic body. The increasing differentiation also expresses itself in prodigious speciation and a wide variety of genera and families, i.e. an ever-growing number of variations upon the basic archetypal theme. There springs to mind in this connection, for instance, the morphological diversity in the Buttercup family, from the Anemone and Marsh Marigold to Columbine and Clematis, Larkspur and Meadow Rue (*Thalictrum*), or the large range of forms found in some fish or insect groups.

The character of this mode of evolution appears especially strikingly in the animal kingdom. Here are found not only tendencies towards structural diversity, but often gradual increases in size as well, extending to gigantism (later Ammonites, Saurians, among birds the Moas, numerous, largely extinct mammals) and, associated with it, over-development of certain parts of the body: the sutures of Ammonites, Elephant tusks, the great, forked nose-parts of the extinct Titanotheres, the antlers of the Great Elk, etc. All these are clear instances of "orthogenesis" (*cf.* Schindewolf 1950)[7], i.e. evolutionary processes that follow a clearly recognisable and definite line. This begins from simple, primeval, small and often structurally indeterminate forms, and soon enters a flourishing phase of prolific diversity. It then proceeds, however, to drain to the limit all available formative potential, which results in over-development and obsolescence, followed inevitably by extinction. The pattern is unmistakable: youth, maturity, old age, senility, death.

In contrast to anagenesis there is another evolutionary tendency which works, as it were, the other way round. It will here be termed *juvenilisation*. It presents

Figure 12: Monocotyledons have (with the exception of the Arum Family) simple, "primitive" leaves, while at the same time – in the Orchid family – displaying within their ranks the most complex flowers of the whole plant kingdom. From left to right: Paradise Lily, (*Paradisea*), Woodrush (*Luzula*), Daffodil (*Narcissus*), Ramsons (*Allium arsinum*). (After Hess, Landolt, Hirzel 1967.)

itself to us with particular clarity in the plant kingdom, most strikingly perhaps in the above-described divergence between trees (or, more generally, woody plants) and herbaceous plants. Development along orthogenetic lines is strongly in evidence here. There is broad agreement today that, of the two, the tree is the more primeval form of life, while its herbaceous counterpart is the product of a later, divergent stage. Takhtajan (1973)[8], for instance, points out that "the secondary nature of the herbaceous form of growth is demonstrated by numerous facts from comparative morphology and systematics. While herbaceous forms are very rare among the *Magnoliales* and *Laurales* (Magnolia and Laurel families — all archaic, angiospermic flowering plants), towards the top of the phylogenetic tree they begin to dominate, becoming most prevalent in the sympetalous families ... It would appear that as a rule herbaceous plants have a more progressive structure than their woody relatives. Finally let it be remembered that among both the recent and the extinct Gymnosperms herbaceous plants are unknown". Recent here refers chiefly to Conifers (they have an ancestry going back much further than the relatively young Angiosperms, to which all herbaceous plants belong).

That which was termed "progressive" in the passage just quoted requires looking at more exactly. If we consider only the vegetative level of the plant, leaving the flower out of account for the time being, it is clear that where "juvenilisation" is concerned there is no question of "anagenesis", of "progressive" evolution in the sense of increasing differentiation, speciation and functional interdependence (*cf.* Remane 1952)[9], or of a "subordination of the parts" (Goethe, *cf.* p 55) to a higher principle regulating the whole organism. Quite the contrary is the case. In herbaceous plants, unlike their woody counterparts, it is precisely the ontogenetic recapitulation of anagenesis, "ascending" leaf by leaf from a simple, archaic structure to one fully differentiated into blade and stalk, that is suppressed or missed out. This highest level is where the herbaceous development actually begins, and, where flowering is involved, leads in the reverse direction to structures that increasingly forego differentiation. This particular line of development is

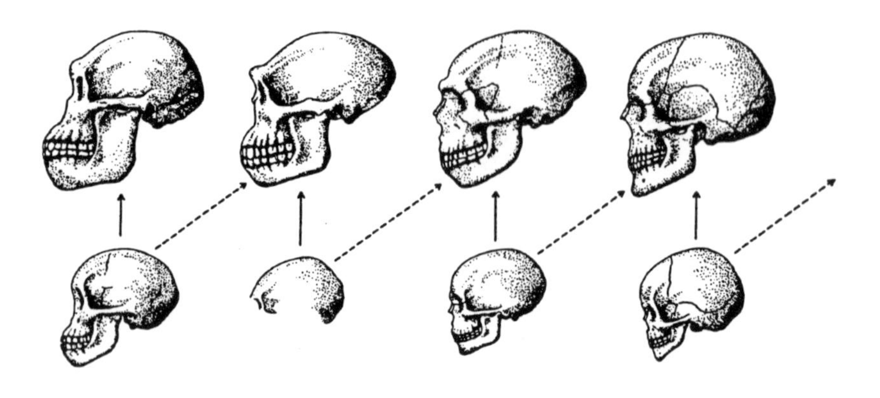

Figure 13: Increasing "juvenilisation" of the Hominids, shown by comparing juvenile (below) with mature stages (above) of, from left to right, *Australopithecus, Homo Erectus, Homo neanderthalensis* and *H. sapiens*. Each subsequent form remains (more or less) at the juvenile stage of its antecedent (after Schindewolf 1972).

something that the herbaceous plant has made very much its own. In the woody plant, by comparison, it remains much more rudimentary.

Within the angiospermic flowering plants this tendency is carried further. Most Monocotyledons, in other words Lilies and their relatives, Grasses, Orchids, go through no more than the most juvenile of these evolutionary stages, namely that of shooting, thus even skipping the preceding phase of regression from "old" to "young" leaf, which is still faithfully followed by the Dicotyledons. To quote Takhtajan once more: "In comparison to most Dicotyledons, the typical Monocotyledons (including all primitive forms) are characterised by a certain "infantilisation" in the vegetative realm. The sort of simplifications they display look like products of a truncated ontogenesis. Their cambium is reduced to an axial activity, and the main root does not develop. The leaves either remain completely undifferentiated or divide indistinctly into stalk and blade, and resemble in their venation the incompletely developed leaf-organs of Dicotyledons (stipules, bracts, bud-scales, sepals etc.)*(Figure 12)*. Taken together these facts have led me to suppose that neoteny (= foetalisation: the preservation of juvenile traits in maturity) has played a decisive role in the origin of the Monocotyledons ... Neotenous development is, to my way of thinking, the key to understanding their morphological peculiarities."

Evolution in both the plant and the animal kingdom is every bit as juvenilistic as it is "progressive". In animals juvenilisation occurs along the most varied lines (*cf.* Schad 1989 for an account of it in an extinct Cephalopode group, the Ceratites)[10]. It is, for instance, a fundamental feature of the evolution of the mammalian skeleton. In this process the primeval, armour-like exoskeleton is gradually reduced, or rather its formation is interrupted at ever more juvenile, rudimentary stages, until all that is left of the exoskeletal tendency is the skull; at the same time the process creates an interior opening for the newly-forming endoskeleton (Suchantke 1983)[11]. At the tip of this great evolutionary line this motif then appears, renewed and enhanced, in man. Through the numerous finds of hominid remains of the last few decades we are well informed about the stream of juvenilisation that has led step by step to modern man *(Figure 13)*. This "foetalisation" is well-known as a basic factor in human evolution. It seems unnecessary to offer yet another description of this, as it has already been done so often in the literature[12].

The Reversal of the Biogenetic Law and the Nature of the Blossom

By means of various examples we have managed to show that juvenilisation occurs, but for this very reason we must now turn to the question of what developmental tendencies these phenomena express — what is the "motif" of this reversal of the Biogenetic Law? In contrast to anagenetic development, which favours the formation of bodily structures, the complete flow of morphic potential into the physical body, in which it becomes completely bound and literally "smothered", juvenilisation signifies a return to the origin, in the sense of a state in which all possibilities are still (or once again) open.

It is certainly not a regression to some past state, but through it the organism is able to begin something new on a different level, or is primed for such a development. We have just seen that there are parallel evolutionary tendencies in man and flowering plant, and a comparison between these is particularly revealing. It shows in each case that the process of juvenilisation is a prelude to a genuine *leap in evolution, i.e. the emergence of something new and completely unprecedented.**

This is especially clear in the case of the plant. It makes the leap from the lower to the higher level in the transition from vegetative leaves to corolla. The latter really does represent something differing utterly in quality from the vegetative realm below. While both green leaves and petals share the same basis as regards organic material, and therefore the same essential leaf-nature, it is nonetheless true that the structures in the corolla are worked upon by formative impulses of a thoroughly different nature.

Accordingly, upward growth stops, and whereas individual organs had formerly been positioned one above the other in spiral formation up the stem, they now group themselves in concentric circles — on the outermost one the sepals, then the petals, the stamens, and finally the stigmata. It can be observed how this tendency has gradually taken hold in the evolution of the flowering plants. While the most primitive type of blossoms, although ostensibly compacted, are still spiral in structure (Magnolias), this "encompassing" tendency ultimately reaches beyond the individual blossom and unites whole groups of them into a flower of a higher order (the pseudanthium of the Compositae, i.e. Daisies, etc.): (wayward) multiplicity conforming to (strictly ordered) unity. From this it is evident that the blossom is not simply an organ in the usual sense, but the product of a morphic *field*, which subordinates to itself any organs that fall within its range. That the formative activities at work in the blossom do not stem from individual organs is shown by the fact that a wide variety of structures can be incorporated into the sphere of the corolla and take on the form and colour of petals: e.g. sepals (Christmas Rose), upper leaves (*Bougainvillea, Poinsettia*), or — most commonly — anthers (in the breeding of ornamental flowers man has taken hold of this natural tendency to turn rudimentary anthers into petals, and, as it were, "artistically" intensified it, e.g. *Paeonia*).

These examples clearly show, on the one hand, at what level the flowering and vegetative regions inter-relate, and on the other, where novelty comes into play. The former refers to the physical nature of the organs, or their rudiments, which are taken up from the vegetative realm into the blossom. These "leaves" will either be stipules, sepals, stamens, and the like, or the spirally-growing organs of the magnolian-type flower-axis. It is the transforming agent, however, as it then works upon this organic material and forms it into a blossom, which brings the novel, the original, into the picture. This reveals itself in all the aspects of the flower that we find directly appealing and moving — its form, colour, scent — everything, in fact, by which the flower expresses itself. These, of course, are all

---

* Translator's note: It might be thought that there is a contradiction between this and the statement in the "Proteus" article (p.15) that "there is ultimately nothing at either end of the plant which is so ... totally new that it was not already present in the leaf". What is being called unprecedented in the present case is the quality whereby the blossom attains to the same ontological level as that of human inner life. In morphological terms it is as much "leaf" as ever.

Figure 14: Dismantled blossom of a garden Tulip. The three inner petals have regressed to a transitional stage between stamen and petal. In other members of the Lily family, e.g. Garlics (*Allium*), such widened, wing-like stamens are quite normal.

phenomena belonging to the realm of sensibility. And where something is expressed, this as a rule implies some opposite number equipped with the requisite senses and sensibility for being appropriately *impressed*. The sentiments, or sensations, which thus arise are, so to speak, the inner equivalent of that which appears in the form and colour of the flower as an *image*.

Blossoms are not organs for expressing the plant's own feelings and sensations. It has none; for feelings come and go and change, while the outward image of the blossom remains fixed (unlike the facial expressions of man and animal). Above all, however, feelings — in spite of being dream-like — represent conscious activity, and are therefore bound to the physical correlative of consciousness, the nervous system, bound, in other words, to the bodily organs. Plants do not have a nervous system. Nevertheless, the forms, colours and scents of flowers are so closely bound up with human sensibility (and on a lower level with animal instinct as well) that they have long been experienced as a version of it existing, as it were, on another plane of being. The truth of this comes across in our love towards flowers, in many of their traditional names, and in the fact that they are used to express our feelings in the form of gifts. There is, however, a further dimension to this. While encounters with animals provoke in us a strong reaction either of sympathy or antipathy (the former for young animals, the latter for spiders and snakes), the sentiments awakened in us by flowers are, as a rule, much less emotionally vehement. They are more subtle, "noble", and belong to a level of inner experience of a stiller, purer and more refined nature. This itself is a sentiment of long standing, and comes to particularly poignant expression in Renaissance paintings of the Madonna. The Lily, as it appears in pictures of the Annunciation is a representation of innocent purity. After the birth of the Boy Jesus we find Mary depicted in a bower of Roses. At the same time, in the image of the Rose reference is made to the deepest of Christian mysteries, that of inner transformation and resurrection: the tangle of thorns as an expression of death, but also of pain, suffering, the repressed forces of the lower nature. The transformation of the thorns is accomplished through contact with the cosmic power of light, through which the blossom arises. Its purity thus represents the power of the inner light of love to transform the lower nature of man such that he develops higher culture-shaping faculties. The equivalent of this in Eastern cultures is the Lotus-blossom, another natural symbol whose eloquence is, if anything, still more immediately comprehensible: delicately pink-tinted, radiant white, and set high upon a slender stalk, it holds sway over the crocodile-inhabited swamp in which it spreads its roots and sets new growth.

The inter-connections involved here become even more impressive if we recall what was said previously about petals; namely, that in most cases (including the Roses, Lilies and Water-Lilies, to which the Lotus belongs) they develop out of rudimentary anthers. Anthers are nothing other than reproductive organs. Their transformation into petals appeals to our sense of beauty and awakens in us feelings of admiration. This must surely mean that we experience directly the deeper dimensions of this developmental process, and that in turn its essential nature finds a correspondent resonance in our inner life. It may also be pointed out in this connection, that man takes cultural advantage of this tendency, and indeed enhances it, when he takes plants out of the wild for the purposes of breeding ornamental varieties.

With this the point has already been made as regards the nature of the new phase of human evolution, for which man has been predisposed by the action of juvenilisation on his body. It has to do with the inner correspondence to that which the plant, in the physical form of its blossom, presents as an outward image; in other words, the plant experienced inwardly as a representation of how the lower impulses of our inner nature can be transformed into higher faculties. *This is yet again analogous to flower-formation in the plant kingdom, in that it is not simply a linear continuation or mere intensification of something already begun. The necessary rudiments are indeed available, but they must be taken hold of by something new, higher, and lifted up to its level. Human individuality, the unique indwelling self, provides, in its aspirations and strivings, the only source of such novelty.* Thus begins the "new" evolution, and that under a thoroughly different emblem from the "old" evolution that came before it. Development used to be regulated according to laws to which the individual unconsciously submitted, being thus entirely at their mercy. Now, however, the Self is gradually taking their place, and assuming control over the lower forces of nature and the responsibility for their transformation. That which was anterior becomes interior, something previously acted upon becomes self-activating. It must be admitted that this new phase of evolution is still in a fluid state, still, indeed, at its very beginnings — not unlike that of the blossom.

"Seekest thou what is highest, greatest? Look to the plant. What it is by nature be thou by intention. Look no further!" It fell to the genius of Friedrich Schiller to capture this parallel, this "unity in diversity", in the most succinct terms. (Friedrich Schiller, *Epigramme*.)

The Two Antagonistic Time-Streams

In conclusion it remains to return to our starting point and cast a glance at the directionally-opposed morphic movements as they appear in the plant at the level of organogenesis, and that of the unfolding of the whole plant in the course of metamorphosis.

Before proceeding, there is a certain fact involved here which it would be useful to explain, as it might otherwise cause misunderstanding. It has been emphasised that the blossom, with its unprecedented qualities, is not just some sort of linear continuation or mere intensification of that which was begun at the vegetative level by metamorphosis. In spite of this the contrary impression might have arisen, for the "metamorphosis" is so closely associated with the process of flowering that it actually only appears on those shoots which are preparing to do so.

The temporal relations between the blossom and the metamorphosis are such that the blossom, although the later phenomenon, and not derivable from what precedes it, is nevertheless that which makes sense of the earlier phase. *The blossom exerts its influence from the future, thus preparing* (by means of the metamorphosis) *its physical emergence.* Starting from the still unformed blossom and viewing the process in *reverse*, we can see in the gradual holding back of vegetative structure how the new morphic process spreads its influence. The resulting store of unused morphic potential then stands fully at the disposal of the blossom, whatever new formative gesture this entails.

Wherever it appears in evolution, juvenilisation is the embodiment of a time-stream which runs counter to one coming from the past responsible for all maturation and ageing processes. It acts not only in the individual plant, but also, as has been shown, in the general evolution of the flowering plants. It is, as it were, the driving force behind them; behind the development, as previously described, from woody to herbaceous forms, as well as that within the flowering plants from Dicotyledons to Monocotyledons.

The honour of being the first to have drawn attention to the relationship involved here belongs to Gerbert Grohmann. In 1931 he speaks of a "Biogenetic Law which encompasses not the past, but the future" [13]. And in a later-published work, "The Plant" [14], he makes the following remarks in connection with the emergence of flower-like structures in Carboniferous Seed Ferns long before that of true flowering plants: "The gesture of a seed-bearing plant is clearly present, even though the actual process involved is still thoroughly that of a fern. Thus, among the ferns we come up against phenomena *which can only be explained in relation to later evolutionary periods"*. And further on: " ... we are forever encountering evolutionary facts, which are only explicable as repetitions of phylogeny. Accordingly the Biogenetic Law points back into the past ... It is a different matter when, following suggestions made by Rudolf Steiner, we, *as it were, reverse the Biogenetic Law,* so that it takes in not just the past, but the future as well. *Rudolf Steiner made the discovery, surprising as it may seem to our normal way of thinking, that it is possible for a developmental phase not only to repeat something previous to it, but also to prefigure later stages in a rudimentary way"* [15].

That there are two opposed time-streams is something that is thoroughly familiar to us from our inner life. We constitute the perpetual meeting point of these two streams. On the one hand, we are moulded and conditioned by the events and experiences of the past; on the other hand, future events cast their shadows "from beyond", or more concretely, future aims or demands determine how we feel and act just as much as those of the past. The projected but still unformed future participates as idea in the reality of the present. In contrast to the relation we bear to our past, with regard to the future we are essentially free agents — insofar, at least, as we are aware of this. For the animal the situation is different. Here morphic potential is devoted entirely to the design and formation of the body. The animal is thereby more highly developed physically, "more mature", but at the same time older with "less of a future" than man. Juvenilisation in man, as in the plant, means the "reserving" of this potential, which, since not exhausted in the structuring of the body, remains at the disposal of the new, higher phase of evolution, which is an inner process. Similar to the plant again is the fact that the physical juvenilisation is only the prelude to the actual process, which has its active source in the future, and comes to fruition through using the available morphic potential. As has been stressed before, this process involves individuality, the Self, which expresses itself in becoming aware of its own nature, and realises itself through its moral, personal and cultural striving.

It seems we must become accustomed to the idea that here we have to do with events in which not only man is involved, but nature as well. With this decisive difference, however: that which in nature is "unwilled" process, going on "by nature", as Schiller has it, is entrusted in man to the conscious will. This does not simply mean that man is capable of dealing with nature as he sees fit, but that he also has, whether he likes it or not, taken over responsibility for the further course of evolution. It depends upon him whether the processes of senescence and obsolescence in nature, the effects of the "old" evolution, will continue to increase one-sidedly, or whether the "young" stream flowing in from the future will gather strength.

NOTES :

1 W. Troll: *Vergleichende Morphologie der höherer Pflanzen*. Berlin 1937-1943 – Die Inflorenszenzen. Stuttgart 1964, 1969.
2 G. Heberer: *Die Theorie der additiven Typogenese*, in G. Heberer (Ed.): Die Evolution der Organismen Vol II/1, 3 ed. 1974.
3 Unpub. manuscript, student project at the Institut für Waldorfpädagogik, Witten (Germany).
4 This and the following quotations from W. Zimmermann: *Geschichte der Pflanzen*, 2nd ed. Stuttgart 1969.
5 J.W. Goethe: *Zur Morphologie*. 1817.
6 A. Suchantke: *Der Kontinent der Kolibris – Landschaften und Lebensformen in den Tropen Südamerikas*, p.134. Stuttgart 1982. W. Schad: *Die Zeitgestalt in der Evolution der Ceratites-Ammoniten aus dem Oberen Muschelkalk Mitteleuropas*, p.133: *Der Doppelstrom der Zeit*, in W. Arnold (Ed.): Entwicklung - Interdisziplinäre Aspekte zur Evolutionsfrage. Stuttgart 1989.
7 O. Schindewolf: *Grundfragen der Paläontologie*. Stuttgart 1950.
8 A. Takhtajan: *Evolutionary Trends in Flowering Plants*. New York 1991.
9 A. Remane: *Die Grundlagen des Natürlichen Systems, der Vergleichenden Anatomie und der Phylogenetik*. Leipzig 1952.
10 W. Schad, see footnote 6.
11 A. Suchantke: *Konvergente Evolution des Skelettes in verschiedenen Tiergruppen*, in W. Schad (Ed.): Goetheanistische Naturwissenschaft Bd. 3: Zoologie. Stuttgart 1983.
12 An up-to-date overview gives the book of A. Montagu: *Growing Young*. New York 1981.
13 G. Grohmann: *Entwicklungsgesetze in der fossilen Pflanzenwelt*, in Gäa Sophia, Jahrbuch der Naturwiss. Sektion der Freien Hochschule am Goetheanum Dornach 1931.
14 English translation: G. Grohmann: *The Plant*. Rudolf Steiner Press, London, 1974.
15 Unfortunately Grohmann gives no reference here. In spite of the efforts made by staff of the Rudolf Steiner Archive at the Goetheanum – to whom hearty thanks are due – the authenticity of the statement could not be established. Steiner does repeatedly refer to a "reversed Biogenetic Law", but always in relation to human biography: *Die Geschichte der Menschheit im Lichte der Geisteswissenschaft* (13.3.1920), Die Menschenschule 10 (1936), No. 5/6, p. 155. Answer to a question concerning the lecture "Anthroposophie und gegenwärtige Wissenschaften" in: Geisteswissenschaft und die Lebensforderungen der Gegenwart, Dornach 1950. Remarks in a debate on psychiatry (26.3.1920), in: Physiologisch-Therapeutisches auf Grundlage der Geisteswissenschaft, Dornach 1975 (GA 314). Nevertheless, Grohmann is perfectly correct (as the present paper tries, if anything, to show) in extending the validity of Rudolf Steiner's observation to the kingdoms of nature, in particular to the plant world. Moreover, for anyone wishing further to pursue the parallels between plant and human evolution, the works of Steiner referred to will help illuminate connections not considered here.

Notes on the Authors

Jochen Bockemühl — born in 1928 in Dresden. Studied zoology, botany, chemistry and geology in Dresden and Tübingen. Since 1971 he has been leader of the Nature/Science section of the laboratory at the Goetheanum in Dornach, Switzerland.

Andreas Suchantke — born in 1933 in Basel, Switzerland. After studying zoology and botany he taught at the Rudolf Steiner School, Zürich, and worked for several years in teacher training courses at Witten, Germany and in other countries. Besides his interests in morphological questions he is engaged in ecological and biogeographical research in tropical Africa, South America and Israel. He has written many books and articles on ecological themes and on environmental education.